Birkhäuser

CIMAT Lectures in Mathematical Sciences

This series, published by Birkhäuser in collaboration with CIMAT, reports recent developments in mathematics and its applications, including those in computer science, mathematical statistics, economics, biology, etc., which are close to the research fields cultivated at CIMAT. The texts to be considered will be sets of lecture notes on current or new topics with a strong mathematical content, as well as presentations of classic fields from novel perspectives. The volumes can be the product of specialized lecture courses, advanced summer/winter schools, or collaborative efforts by research groups to present various facets of a given field in an insightful and unified manner.

Gerardo Arizmendi Echegaray •
Luis Hernández-Lamoneda •
Rafael Herrera Guzmán
Editors

Recent Advances in Alexandrov Geometry

Birkhäuser

Editors
Gerardo Arizmendi Echegaray
Actuarial Science, Physics and Math
Universidad de las Américas Puebla
Cholula, Puebla, México

Luis Hernández-Lamoneda
Mathematics
Centro de Investigación en Matemáticas
Guanajuato, México

Rafael Herrera Guzmán
Mathematics
Centro de Investigación en Matemáticas
Guanajuato, México

ISSN 2731-6521 ISSN 2731-653X (electronic)
CIMAT Lectures in Mathematical Sciences
ISBN 978-3-030-99297-2 ISBN 978-3-030-99298-9 (eBook)
https://doi.org/10.1007/978-3-030-99298-9

Mathematics Subject Classification: 53C23, 53C24, 53C20, 57N10, 57S25, 46E36

This book is published under the imprint Birkhäuser, www.birkhauser-science.com by the registered company Springer Nature Switzerland AG
The registered company address is: Gewerbestrasse 11, 6330 Cham, Switzerland

Preface

The Eleventh Mini-Meeting on Differential Geometry was held from December 10th to 12th, 2018, at the Center for Research in Mathematics (CIMAT), Guanajuato, México.

The invited speakers included Peter Petersen (Distinguished Visiting Professor for the Mexican Academy of Sciences and the USA-Mexico Foundation for Science), Fernando Galaz-García, Jesús Angel Núñez-Zimbrón, Gabriel Ruiz, Gregor Weingart, Miguel Angel García-Ariza, Rosemberg Toalá-Enríquez and Armando Cabrera-Pacheco. Peter Petersen and Fernando Galaz-García delivered two advanced mini-courses of three lectures each on Alexandrov Geometry, while the rest of the participants delivered research talks on diverse topics.

This volume is devoted to various aspects of Alexandrov Geometry for those wishing to get a detailed picture of the advances in the field. It contains enhanced versions of the lecture notes of the two mini-courses plus those of one research talk:

- Peter Petersen's notes aim at presenting various rigidity results about Alexandrov spaces (work with Karsten Grove) in a way that facilitates the understanding by a larger audience of geometers of some of the current research in the subject. They contain a brief overview of the fundamental aspects of the theory of Alexandrov spaces with lower curvature bounds, as well as the aforementioned rigidity results with complete proofs.
- The lecture notes of Fernando Galaz-García's mini-course were completed in collaboration with Jesús Nuñez-Zimbrón. They present an up-to-date and panoramic view of the topology and geometry of three-dimensional Alexandrov spaces, including the classification of positively and non-negatively curved spaces and the geometrization theorem. They also present Lie group actions and their topological and equivariant classifications as well as a brief account of results on collapsing Alexandrov spaces.
- Jesús Nuñez-Zimbrón's notes survey two recent developments in the understanding of the topological and geometric rigidity of singular spaces with curvature bounded below.

We thank all the participants for making the meeting a success. We also thank CIMAT's staff for their help in the organization and smooth running of the event. This meeting was the eleventh edition of an annual event intended for researchers and graduate students, with the dual aim of combining a winter school and a research workshop. The meeting was supported by the Mexican Academy of Sciences (AMC), the USA-Mexico Foundation for Science (FUMEC), the Universidad de las Américas Puebla (UDLAP), the Mexican Science and Technology Research Council (CONACyT) and the Center for Research in Mathematics (CIMAT). The organizers were Gerardo Arizmendi (UDLAP, México), Luis Hernandez-Lamoneda (CIMAT, México) and Rafael Herrera (CIMAT, México).

Cholula, México — Gerardo Arizmendi Echegaray
Guanajuato, México — Luis Hernández-Lamoneda
Guanajuato, México — Rafael Herrera Guzmán

Contents

Rigidity of Alexandrov Spaces

Peter Petersen

2000 Mathematics Subject Classification 53C23, 53C24

1 Introduction

These notes are based on three lectures given during the 11th Mini Meeting on Differential Geometry in December of 2018 at CIMAT in Guanajuato. Each of the first three sections corresponds to one of the lectures. The last section contains the main body of work that leads to a complete proof of the most general version of the rigidity results investigated by Karsten Grove and myself. The goal with the lectures were to give a brief overview of the fundamental aspects of the theory of Alexandrov spaces with lower curvature bounds. The hope was that with this background it would be possible for a larger audience of geometers to understand some of the current research in the subject. To assist in the writing of these notes Gerardo Arizmendi typed notes based on the lectures and also added all of the figures in the text, a task for which I am forever grateful. A parallel set of lectures were given by Fernando Galaz Garcia on more topological aspects of Alexandrov spaces, with special attention to his research on the classification of such spaces in lower dimensions. Those results were subsequently used to enhance these notes and completely classify the spaces we were studying in dimensions ≤ 4. A fortuitous outcome of the meeting.

While the theory of Alexandrov spaces has been around for a long time it received a significant boost in modern times with the paper [4]. Soon after, Perel'man distributed the preprint [24], which contained several profound developments and further revolutionized the subject. This preprint was unfortunately never published as Perel'man had hoped to find a different strategy that would lead to better proofs and results. Subsequently, Perel'man and Petrunin significantly enhanced

P. Petersen (✉)
Department of Mathematics UCLA, Los Angeles, CA, USA
e-mail: petersen@math.ucla.edu

G. Arizmendi Echegaray et al. (eds.), *Recent Advances in Alexandrov Geometry*, CIMAT Lectures in Mathematical Sciences,
https://doi.org/10.1007/978-3-030-99298-9_1

the technical tools available for the study of Alexandrov spaces. The most complete reference guides that explain all of this material are [18, 27, 28]. These papers, together, also contain complete proofs of all of the significant results. However, the uninitiated reader might want to start by consulting [3] before digging deeper into the subject.

The first two sections of these notes contain an outline that includes most of the crucial concepts and results that can be found in [18, 27, 28]. There are essentially no proofs in these sections. The exceptions in Sect. 3 came about as I realized that certain basic aspects appeared mysterious to several people. The last two sections explain the questions that Karsten Grove and I have been working on including complete proofs. The proofs use all of the material from the first two sections as well as several recent papers.

It is a pleasure to thank Gerardo Arizmendi, Rafael Herrera, and Luis Hernandez who organized the meeting and procured the necessary funding and support from CIMAT, CONACYT, Academia Mexicana de Ciencias, FUMEC, UDLAP, and Sociedad Matemática Mexicana. I would also like to thank Adam Moreno, Matthias Wink and the referees who read through the notes and provided useful feedback that significantly improved the text.

2 The Basic Theory

In the first section we survey the main properties of Alexandrov spaces with lower curvature bounds and also offer several examples. The material in this section was originally developed in [4] and is nicely explained in detail in [3], one of the most accessible sources for the elementary aspects of Alexandrov spaces. A richer more in-depth treatment can be found in [28].

2.1 Spaces and Geodesics

We start with a metric space $(X, |\cdot\cdot|)$, where $|pq| = \text{distance}\,(p, q)$. The first basic assumption is that all closed balls $\bar{B}\,(p, r) = \{x \in X \mid |px| \leq r\}$ are compact. It is also traditional to assume that the space has finite dimension. Here the most convenient assumption is to work with finite Hausdorff dimension. This will be also be convenient for defining volume and obtain Gromov-Hausdorff precompactness.

The metrics are also required to allow for minimal geodesics, also called segments, between all pairs of points:

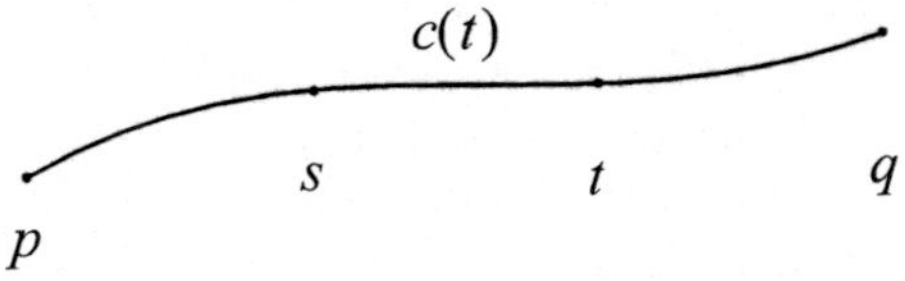

These are curves $c : [a, b] \to X$ with the property that: $|c(s)c(t)| = L|s - t|$ for all $s, t \in [a, b]$. Here L is the speed of the curve and we often assume that: $L = 1$. In fact one can easily reparametrize c to become such a unit speed geodesic $\bar{c} : [0, |pq|] \to X$ with $|\bar{c}(s)\bar{c}(t)| = |s - t|$. If $c(a) = p$ and $c(b) = q$, then we say that c is a geodesic from p to q. We will assume that such geodesics exist between all pairs of points. It is occasionally convenient to use the notation $\overline{pq}$ for a particular choice of segment between p and q.

A somewhat weaker looking assumption is often used. The approximate midpoint property assumes that for all $p, q \in X$ and $\epsilon > 0$ there exist m such that $|mp| + |mq| \leq |pq| + \epsilon$. Using that the space is complete one can find midpoints where $\epsilon = 0$, it is then easy to iterate this procedure to define the potential geodesic on dyadics, i.e., points at half, quarter, eighths, etc. distance between p, q. This defines a uniformly continuous curve on a dense subset of $[0, |pq|]$ and the geodesic is the unique extension of this curve.

We shall also work with Lipschitz curves:

$$c : [a, b] \to X, \text{ where } |c(s)c(t)| \leq L(s - t).$$

For such curves the length is given by

$$L(c) = \sup_{a=t_0<t_1\cdots<t_n=b} \sum_{i=0}^{n-1} |c(t_i)c(t_{i+1})| = \int_a^b s(t)dt,$$

where $s(t_0)$ is the speed of the curve defined by

$$s(t_0) = \lim_{t\to t_0^+} \frac{|c(t_0)c(t)|}{t - t_0}.$$

The speed exists for almost all t_0. However, showing that

$$\sup_{a=t_0<t_1\cdots<t_n=b} \sum_{i=0}^{n-1} |c(t_i)c(t_{i+1})| = \int_a^b s(t)dt$$

is not entirely trivial (see [3]).

2.2 Lower Curvature Bounds

Next we define what it means for such a metric space to have curvature $\geq k$, i.e., be an Alexandrov space with a lower curvature bound.

This is done by using the complete simply connected 2-dimensional space forms S_k^2 as *comparison spaces*. The n-dimensional space forms S_k^n are the round spheres S^n of radius $1/\sqrt{k}$ for $k > 0$, Euclidean space $\mathbb{R}^n$ for $k = 0$, and hyperbolic space scaled by $1/\sqrt{-k}$ for $k < 0$.

Let $\boldsymbol{\Delta(p, q, r)}$ be a triangle of segments between the three points in $(X, |\cdot\cdot|)$ and $\boldsymbol{\Delta(\tilde{p}, \tilde{q}, \tilde{r})}$ a comparison triangle on S_k^2, i.e., $|pq| = |\tilde{p}\tilde{q}|$, $|qr| = |\tilde{q}\tilde{r}|$, and $|pr| = |\tilde{p}\tilde{r}|$. We say that the triangle $\boldsymbol{\Delta(p, q, r)}$ has curv $\geq k$ provided: for any points q' and r' on the segments $\overline{pq}$ and $\overline{pr}$, respectively, and points $\tilde{q}'$ and $\tilde{r}'$ in the segments $\overline{\tilde{p}\tilde{q}}$ and $\overline{\tilde{p}\tilde{r}}$, respectively, with the properties that $|pq'| = |\tilde{p}\tilde{q}'|$ and $|pr'| = |\tilde{p}\tilde{r}'|$ it follows that $|q'r'| \geq |\tilde{q}'\tilde{r}'|$

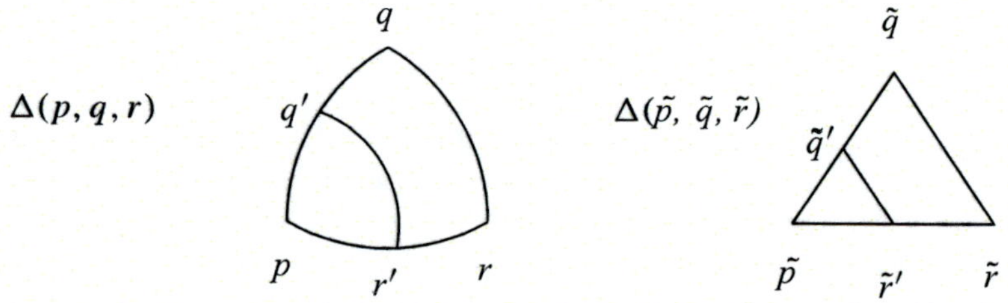

When $k > 0$ the comparison space is clearly compact. Thus it is necessary to ensure that the comparison triangle exists.

Definition 2.1 **With that in mind we define a complete metric space** X to be an n-dimensional Alexandrov space with curv $\geq k$ provided the Hausdorff dimension is n, all pairs of points are joined by minimal geodesics, and every point has a neighborhood U such that any triangle in U has curvature $\geq k$.

The globalization theorem for Alexandrov spaces (see [3, 4]) implies that this **local definition leads to global comparison for all triangles, i.e., any triangle in** X admits a comparison triangle and has curv $\geq k$. This is also known as Toponogov's theorem as he proved this global result for Riemannian manifolds.

Since curvature comparison is more or less trivial for spaces of dimensions zero and one, we need to decree what type of comparison should be allowed. As for dimension zero it is reasonable to assume that we only consider one point spaces. However, it will be convenient to also allow for the disconnected space $S^0 = \{\pm 1\} \subset \mathbb{R}$ with the understanding that the two points are distance π apart. Both the one point space and the two point space $\{0, \pi\}$ are considered to have curv ≥ 1. In dimension one all circles and intervals, whether open, closed, or half open, always have curv ≥ 0. Moreover, when circles have length $\leq 2\pi$ and the intervals length $\leq \pi$ they have curv ≥ 1. Thus circles, closed intervals, and $\mathbb{R}$ are the only one dimensional Alexandrov spaces.

2.3 Angles

To define angles between geodesics we begin by defining the preliminary angle $\measuredangle_k(qpr)$ at p as the angle at $\tilde{p}$ for the comparison triangle $\Delta(\tilde{q}, \tilde{p}, \tilde{r})$ in S_k^2 . This is well defined as long as q, r are sufficiently close to p. The angle can be calculated using the distances between p, q, r and the law of cosines in the space form. In an Alexandrov space we define the angle between two geodesics with $c_1(0) = c_2(0) = p$ by

$$\measuredangle(c_1, c_2) = \lim_{s,t\to 0} \measuredangle_k(c_1(s)pc_2(t)).$$

Here it is natural to use k as the lower curvature bound so as to obtain the angle as a monotone limit:

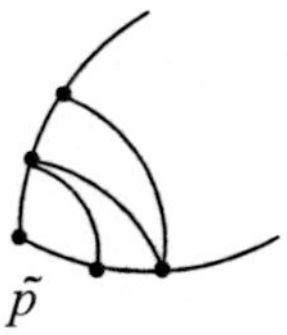

Monotonicity follows from triangle comparison and it is not hard to see that the limit is the same for any choice of k as the model geometries agree up to first order. This metric angle was first introduced by Alexandrov.

The law of cosines in constant curvature k shows that the distance comparison above is equivalent to the following version of distance comparison: Consider unit speed geodesics c_1, c_2 emanating from $p \in X$ and unit speed geodesics $\tilde{c}_1, \tilde{c}_2$ emanating from $\tilde{p} \in S_k^n$. If $\measuredangle(c_1, c_2) = \measuredangle(\tilde{c}_1, \tilde{c}_2)$, then $|\tilde{c}_1(s)\tilde{c}_2(t)| \leq |c_1(s)c_2(t)|$ for all s, t:

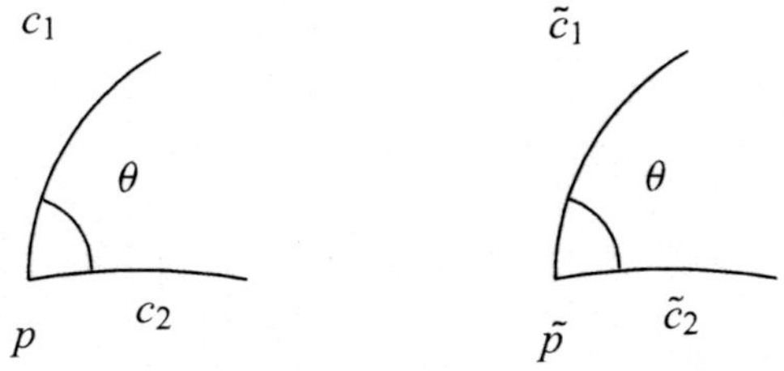

The distance comparison just defined can easily be seen to be equivalent to the more elegant angle comparison condition.

$$\measuredangle(pqr) \geq \measuredangle_k(pqr),$$

where $\measuredangle(pqr)$ is the smallest angle $\measuredangle(c_1, c_2)$ for segments from q to p and r.

Angles also satisfy the properties:

$$\measuredangle(c_1, c_3) \leq \measuredangle(c_1, c_2) + \measuredangle(c_2, c_3)$$

$$\measuredangle(c_1, c_2) + \measuredangle(c_2, c_3) + \measuredangle(c_3, c_1) \leq 2\pi$$

2.4 Examples

Below we give a few basic examples of Alexandrov spaces.

Example 2.2 Any closed convex subset $C \subset X$ of an Alexandrov space is itself an Alexandrov space. Here convexity can be defined by declaring that for each pair of points in C, there exists a geodesic between them in X that also lies in C. One could change this to include all possible geodesics between the points in X (total convexity), or for L-convexity just all geodesics of length $< L$. Note that on $S^2(1)$ a pair of antipodal points form a π-convex set, but not a convex set. The equator is a convex set that is also π-convex, but it is not totally convex.

Example 2.3 ([4j]) The *Euclidean cone* over a general metric space X is $C_0(X) = \{(r, x) | r \geq 0, x \in X\}/\{0\} \times X = \text{point}$

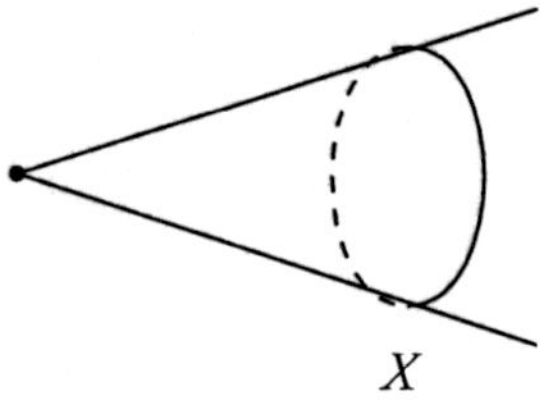

with distances calculated according to the Euclidean law of cosines

$$|(r_1, x_1), (r_2, x_2)|^2 = r_1^2 + r_2^2 - 2r_1 r_2 \cos(|x_1 x_2|_X).$$

Here $|x_1 x_2|_X$ represents the angle between the curves $t \mapsto (t, x_i)$ at the vertex in $C_0(X)$. Note that when $\theta(t)$ is a geodesic of length $\leq 2\pi$ on X, then $C_0(\theta(t)) \subset C_0(X)$ is a flat cone that can be cut out of $\mathbb{R}^2$.

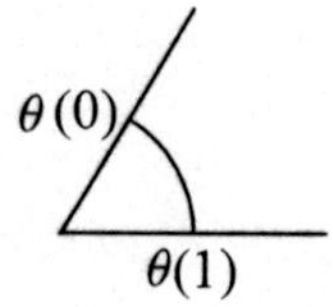

Similarly, there are cones $C_k(X)$ which are defined using the law of cosines in S_k^2. When $k > 0$ the cone is restricted to have $r \in \left[0, \frac{\pi}{2\sqrt{k}}\right]$ so as to remain convex. For $k = 1$ distances are calculated using

$$\cos|(r_1, x_1), (r_2, x_2)| = \cos r_1 \cos r_2 + \sin r_1 \sin r_2 \cos(|x_1 x_2|)$$

and when $k = -1$ via

$$\cosh|(r_1, x_1), (r_2, x_2)| = \cosh r_1 \cosh r_2 - \sinh r_1 \sinh r_2 \cos(|x_1 x_2|).$$

It is not hard to see that when X is an Alexandrov space with curv ≥ 1, then $\operatorname{curv}(C_k(X)) \geq k$. More remarkable is the fact that the converse also holds.

Example 2.4 When $(X, |\cdot\cdot|)$ has curv ≥ 1 we can also construct the *spherical suspension* of X by

$$\Sigma_1 X = \{(r, x) | r \in [0, \pi], x \in X\} / (\{0\} \times X = p_1, \{\pi\} \times X = p_2)$$

with distances calculated as in S^2. This space also has curv ≥ 1.

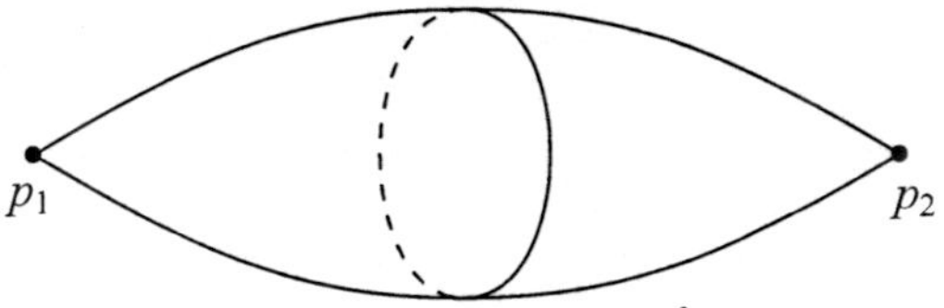

Example 2.5 Given two spaces $(X, |\cdot\cdot|)$ and $(Y, |\cdot\cdot|)$ with curv ≥ 1 we define the *spherical join*

$$X * Y = X \times \left[0, \frac{\pi}{2}\right] \times Y / \left(X \times \{0\} \times Y \sim X, X \times \left\{\frac{\pi}{2}\right\} \times Y \sim Y\right)$$

with distances calculated as in $S^3 = S^1 * S^1$. Explicitly:

$$\cos|(x_1, r_1, y_1)(x_2, r_2, y_2)| = \cos(r_1)\cos(r_2)\cos(|x_1 x_2|) + \sin(r_1)\sin(r_2)\cos(|y_1 y_2|).$$

Again $X * Y$ has curv ≥ 1.

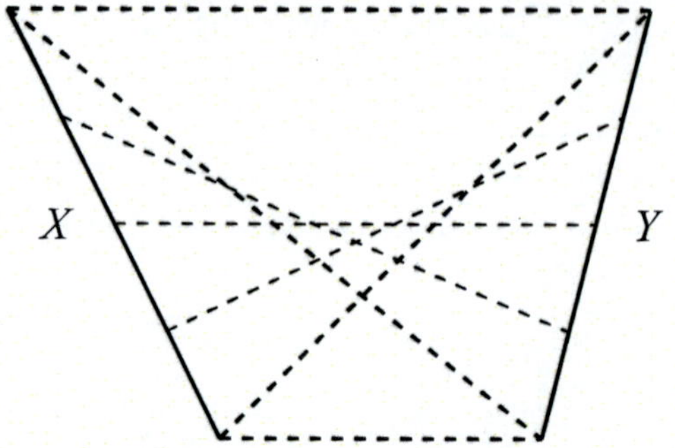

Tetrahedron

Special cases are the spherical suspension: $\Sigma_1 X = \{0, \pi\} * X$ and spherical cone $C_1(X) = \{0\} * X$, which is half of the spherical suspension.

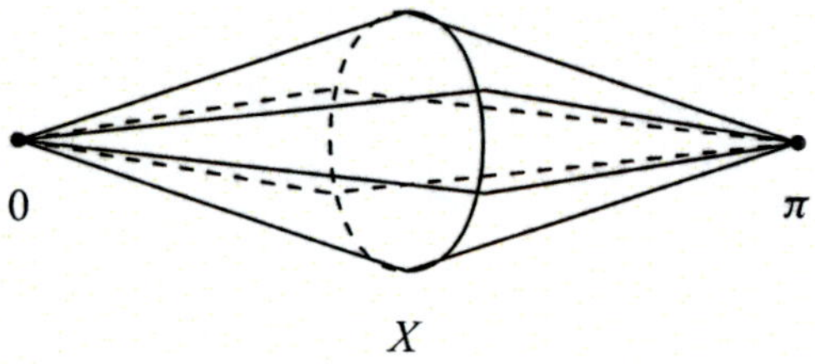

Note that $S^p(1) * S^q(1) = S^{p+q+1}(1)$.

Example 2.6 The *double disk*:

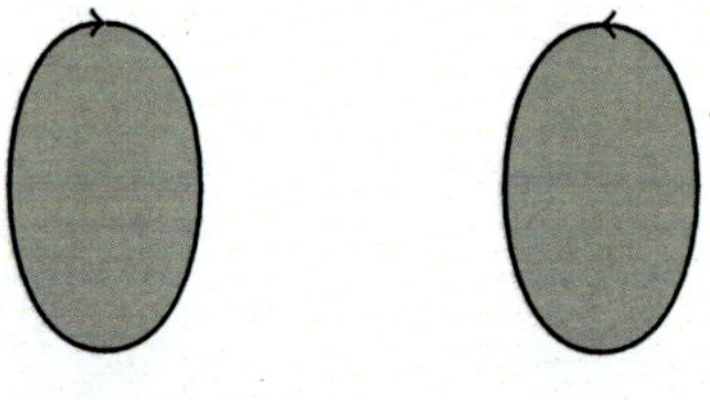

consists of two Euclidean discs glued along the boundary. This space has curv ≥ 0.

More generally, if X is an Alexandrov space with boundary ∂X (to be defined at the end of Sect. 3), then the *double* $D(X) = X \cup_{\partial X} X$ (two copies of X with their boundaries identified) is also an Alexandrov space. The proof of the latter statement is quite involved (see [24]).

Example 2.7 Any Riemannian manifold with sectional curvature $\geq k$ is, by Toponogov's theorem, an Alexandrov space with curv $\geq k$.

Example 2.8 If $(X, |\cdot\cdot|)$ has curv $\geq k$ and G is a group that acts isometrically on X with compact orbits, then X/G has curv $\geq k$. This can be generalized to certain leaf spaces. Consider a Riemannian manifold M with curv $\geq k$ and a *submetry* $f : M \to X$, where X is a metric space. The condition of being a submetry is that for all $p \in M$ and $r > 0$: $f(\bar{B}(p,r)) = \bar{B}(f(p),r)$. In this case X also has curv $\geq k$.

When such a leaf space has the property that the leaves are properly embedded submanifolds in M, then the submetry is also known as a manifold submetry.

Example 2.9 Let $C \subset \mathbb{R}^n$ a compact convex set, then ∂C is also an Alexandrov space with curv ≥ 0.

2.5 Dimension and Volume

The goal of this section is to explain how the dimension enters the picture. The theory of topological and Hausdorff dimension will used and the basic material needed is contained in [16] (this is, in my opinion, possibly the best math book ever written).

The n-dimensional Hausdorff measure of a set $U \subset X$ ($n \geq 0$ is not necessarily an integer) is denoted by $\mu_n(U)$. **The Hausdorff dimension is defined as**

$$\dim_H U = \sup\{n \geq 0 \mid \mu_n(U) = \infty\}.$$

Let $(X, |\cdot\cdot|)$ be an Alexandrov space of curvature ≥ 0. For $r < R$ we wish to create an $\frac{r}{R}$-expanding map $f : \bar{B}(p,R) \to \bar{B}(p,r)$ such that $|f(x)f(y)| \geq \frac{r}{R}|xy|$.

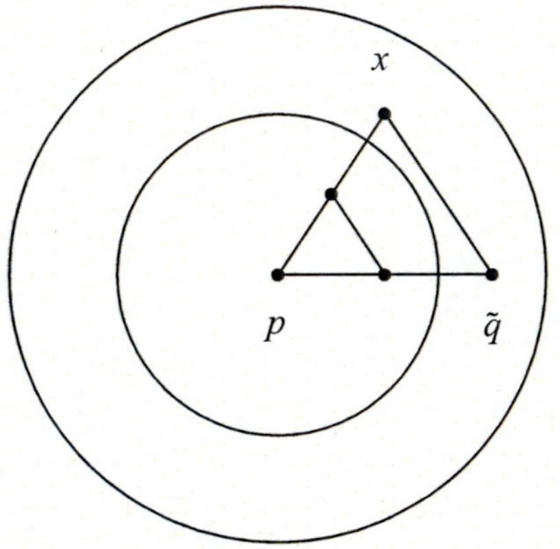

The natural choice is to scale along geodesics: $f(x) = c(\frac{r}{R}|px|)$, with $c(0) = p$, $c(|px|) = x$. For this to make sense we need a way of choosing a geodesic c from x to p. This clearly requires a suitable axiom of choice. As we are interested in measuring sets we need a Borel measurable f. Fortunately, we can invoke the Borel measurable axiom of choice (see [19]). Distance comparison then shows that relative volume comparison holds for any Hausdorff measure:

$$\begin{aligned}\mu_n(\bar{B}(p,r)) &\geq \mu_n(f(\bar{B}(p,R))) \\ &\geq (\frac{r}{R})^n \mu_n(\bar{B}(p,R)).\end{aligned}$$

In particular, if just one ball has finite n-dimensional Hausdorff measure, then all concentric balls have finite n-dimensional Hausdorff measure. This in turn shows that all bounded sets have finite n-dimensional Hausdorff measure. When the lower curvature bound is, say, -1, then we have a similar estimate:

$$|f(x)f(y)| \geq \frac{\sinh(r)}{\sinh(R)}|xy|.$$

Here the inequality is generally strict even in hyperbolic space with $|xp| = |yp| = R$.

One of the important steps in the early parts of the theory of Alexandrov spaces is to show that the Hausdorff dimension is the same as the topological dimension. This can be handled in the following way.

Given an integer $m = 0, 1, 2, \ldots$ we say that $p \in X$ is (m, ϵ)-*strained* if there are m pairs of points $p_i^{\pm}$, $i = 1, \ldots, m$ such that

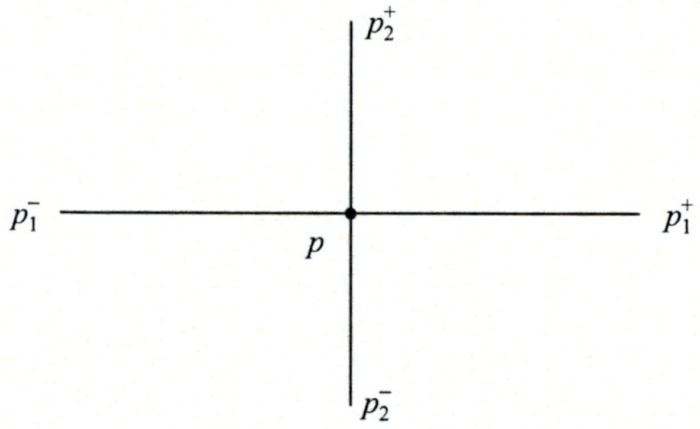

$$\measuredangle(p_i^{\pm} p p_j^{\pm}),\ \measuredangle(p_i^{\pm} p p_j^{\mp}) > \frac{\pi - \epsilon}{2}, i \neq j,$$

$$\measuredangle(p_i^{+} p p_i^{-}) > \pi - \epsilon.$$

It is possible to show that ϵ can be fixed to be a small number that only depends on m. Specifically, there exists $\delta(m) > 0$ such that if a point is (m, δ)-strained, then there are nearby points which are (m, ϵ)-strained for any $\epsilon > 0$. Thus we can simply refer to the point as being m-strained. It is also easy to see that the condition of being (m, ϵ)-strained is an open and even dense condition. The first important thing to prove is that if p is m-strained, then we obtain a map $\phi_m : X \to \mathbb{R}^m$ by

$$x \mapsto \begin{pmatrix} |xp_1^+| \\ \vdots \\ |xp_m^+| \end{pmatrix},$$

which is Lipschitz on X and open in a neighborhood of p. This implies that the strainer number is bounded by the Hausdorff dimension:

$$\dim_H X \geq m.$$

It is also not too hard to check that if this map is not injective in any neighborhood of p, then there are $(m + 1)$-strained points arbitrarily close to p. This suggests a further definition.

The *strainer number* of $p \in X$ is the maximal n such that p is n-strained and there are no $(n + 1)$-strained points arbitrarily close to p. If the strainer number is finite, then the coordinate map $\phi_n : X \to \mathbb{R}^n$ is a homeomorphism onto its image when restricted to a suitable neighborhood around p. This in turn can also be used to check that the map is bi-Lipschitz. We can then conclude that $\dim_H X = \dim X = n$ provided some point p has a finite strainer number n.

This leaves open the possibility that the strainer number can be infinite, while the topological dimension is finite. It is possible with considerable extra effort to show that this is impossible. Specifically, the existence of gradient curves (see next section) for distance functions has to be justified without knowing that the space of directions is a compact inner metric space. The most elementary account can be found in [28], more modern treatments use ultra-filters, which in turn requires the axiom of choice as well as the continuum hypothesis (see [21]). These are very strong assumptions to impose when working with proper metric spaces that are second countable. For Alexandrov spaces the only version of the axiom of choice generally needed is for countable collections of sets.

All in all this tells us that an n-dimensional Alexandrov space has a meaningful Borel measure, the n-dimensional Hausdorff measure, with the property that all bounded open sets have finite nonvanishing measure. We can normalize this measure $\mathrm{vol}_n = c_n \mu_n$ by a universal constant such that it becomes the Lebesgue measure on Euclidean space and the Riemannian volume on Riemannian manifolds. Finally, when $\mathrm{curv} \geq -1$ this measure satisfies the relative volume comparison estimate:

$$\frac{\mathrm{vol}_n B(p, R)}{\mathrm{vol}_n B(p, r)} \leq \frac{\sinh^n R}{\sinh^n r}.$$

One can in fact fairly easily also establish the usual Bishop-Gromov volume comparison where the estimate is the ratio of balls of radius R and r in hyperbolic space.

Example 2.10 Let $\{(X_i, |\cdot\cdot|_i)\}_{i\in\mathbb{N}}$ be spaces with curv $\geq k$. If $\dim X_i \leq n$, then it follows from relative volume comparison that some subsequence will converge in the (pointed) Gromov-Hausdorff sense. The limit also has curv $\geq k$ and dim $\leq n$. Thus pointed Alexandrov spaces with curv $\geq k$ and dim $\leq n$ form a compact class of metric spaces.

This leads to the following interesting construction: Let $(X, |\cdot\cdot|)$ have curv $\geq k$ and scale distances $(X, \lambda|\cdot\cdot|)$ so that now curv $\geq \frac{k}{\lambda^2}$. The accumulation points of the pointed family $(X, \lambda|\cdot\cdot|, p)$ as $\lambda \to \infty$ turn out to be Euclidean cones with curv ≥ 0 and are called the Gromov-Hausdorff tangent cones at p. Note that as these spaces also have the property that closed balls are compact. We shall give a different definition of what a tangent cone is that turns out to be identical to these limits. In particular, the Gromov-Hausdorff tangent cones are actually uniquely defined.

2.6 The Space of Directions and the Tangent Cone

The space of directions at $p \in X$, S_pX, is defined as the the metric completion of the space of unit speed geodesics emanating from p when using the angle as metric. In general a curve c with $c(0) = p$ defines a direction if it has unit speed. This will be its velocity $\dot{c}(0)$ or more accurately $\dot{c}^+(0)$ as we only need to know the curve on one side of p. It will also have a left sided velocity $\dot{c}^-(0)$ if it is defined on the other side. For two such curves the angle between them is the distance between its velocities $|c_1c_2| = \measuredangle(\dot{c}_1, \dot{c}_2)$. It follows from distance comparison that geodesics with angle 0 coincide.

The tangent cone at p is defined as the Euclidean cone over the space of directions: $T_pX = C_0(S_p(X))$. When curv $X \geq k$, it is also often convenient to use the cone $C_k\left(S_pX\right)$ in place of the tangent space.

Example 2.11 Note that $S_{\text{vertex}}C_0(X) = X$ and $S_{\text{vertex}}(C_0(X) \times C_0(Y)) = X * Y$.

By invoking the Borel measurable axiom of choice again we can create:

$$\begin{aligned}\log_p : X &\to T_pX \\ x &\mapsto \left(|xp|, \overrightarrow{px}\right)\end{aligned}$$

In effect, this map assigns a minimal geodesic from x to p for each $x \in X$. When curv $X \geq 0$, it follows from angle comparison that this map is distance nondecreasing. The scaled logarithm $\lambda \log_p(x) = \left(\lambda\,|xp|, \overrightarrow{px}\right)$ is simply the radial scaling of the corresponding tangent vectors. When curv $X \geq k$ and $k < 0$, it is convenient to consider $\log_p : X \to C_kS_pX$ so as to also obtain a distance nondecreasing map. The target cone is the same but we have changed the metric on it.

As mentioned above, the family $(X, \lambda|\cdot\cdot|, p)$ is precompact in the Gromov-Hausdorff topology as $\lambda \to \infty$. The map $\log_p : X \to T_pX$ in fact gives a good approximation as one can show that for all $\epsilon > 0$, there exist $\lambda >> 1$, such that for all $x, y \in \bar{B}\left(p, \lambda^{-1}\right)$:

$$\left|\lambda\,|xy| - \left|\lambda \log_p x\, \lambda \log_p y\right|\right| \leq \epsilon.$$

This is a consequence of how angles were defined as limits of comparison angles. Moreover, as geodesic directions are dense in the space of directions the images $\lambda \log_p \left(B\left(p, \lambda^{-1}\right)\right) \subset B\left(o_p, 1\right)$ will become dense as $\lambda \to \infty$. All in all this shows that

$$\lim_{\lambda\to\infty} (X, \lambda\,|\cdot\cdot|\,, p) - T_pX = C_0\left(S_pX\right).$$

This implies some crucial results: the tangent cones are Alexandrov spaces with $\operatorname{curv} T_pX \geq 0$ and $\dim T_pX = \dim X$. The latter follows from the fact that

$$\dim \lim_{\lambda\to\infty} (X, \lambda\,|\cdot\cdot|\,, p) \leq \dim X,$$

while the nondecreasing map $\log_p : X \to C_k\left(S_pX\right)$ doesn't decrease the Hausdorff dimension. Since the tangent cone is the Euclidean cone over the space of directions we now also get that S_pX is a compact Alexandrov space (T_pX is locally compact) with $\operatorname{curv} \geq 1$ and $\dim(S_pX) = \dim(X) - 1$. These facts for the space of directions, including even the approximate midpoint property, do not seem to have proofs that are more direct or simpler.

A point $p \in X$ is said to be *regular* if the tangent cone is isometric to $\mathbb{R}^n$. This is now evidently the same as saying that p has strainer number n and that p is (n, ϵ)-strained for all $\epsilon > 0$. The Baire category theorem implies that the set of regular points is a dense G_δ set.

A lot of Alexandrov spaces have boundary, e.g., convex sets in Euclidean space are natural examples. The definition of what it means to be a point on the boundary is best understood if we define the boundary inductively by dimension.

Definition 2.12 In dimension 0 we say that the two point space $S^0 = \{0, \pi\}$ has no boundary while the one point space $\{0\}$ does have boundary. For a finite dimensional Alexandrov space X we say that $p \in X$ belongs to the boundary, $p \in \partial X$, when $\partial S_pX \neq \emptyset$. If $p \notin \partial X$, then we say that p lies in the interior of X.

This definition shows that a compact interval $[0, \pi]$ has $\{0, \pi\}$ as boundary and interior $(0, \pi)$. More generally, the closed hemisphere has the equator as boundary.

3 Calculus on Alexandrov Spaces

The goal in this section is to develop some of the more subtle techniques that can be used to investigate Alexandrov spaces. A few of the more elementary aspects of the theory are developed in a little more detail. It is hoped that this will help people appreciate how simple calculus is when it focuses on monotonicity and convexity. The material here is covered in much greater detail in [28] and [27], the latter also contains a proof of the existence of quasi-geodesics. For the proofs of the fibration and stability theorems the reader is probably best served by consulting [18]. A very general and clean account of gradient curves that uses ultra-filters can be found in [21].

3.1 Calculus of Continuous Functions

Consider a continuous function $f : I \to \mathbb{R}$ on an interval $I \subset \mathbb{R}$. Often f is actually Lipschitz as it is the restriction of a distance function to a unit speed curve. We use Calabi's definition of upper bounds for second derivatives that relies on barrier functions.

Definition 3.1 (Calabi) $f''(t_0) \leq \lambda$ if and only if, for all $\epsilon > 0$ there exist α such that

$$f(t) \leq \underbrace{f(t_0) + \alpha(t - t_0) + \frac{\lambda + \epsilon}{2}(t - t_0)^2}_{\text{support of } f \text{ at } t_0}.$$

Note that:

$$f''(t_0) \leq \lambda \iff (f - \frac{\lambda}{2}t^2)''(t_0) \leq 0 \iff (f - \frac{\lambda}{2}(t - t_0)^2)''(t_0) \leq 0.$$

Moreover, $f''(t_0) \leq \lambda$ provided there exists α such that

$$f(t) \leq f(t_0) + \alpha(t - t_0) + \frac{\lambda}{2}(t - t_0)^2 + o((t - t_0))^2.$$

Definition 3.2 If $f : I \to \mathbb{R}$ and $f'' \leq \lambda$ on all of I, then we say that f is semi-concave or λ-concave.

We need the important theorem.

Theorem 3.3 (Minimum Principle) *If $f'' \leq 0$ on $I = [a, b]$ and $f(t) \geq f(t_0)$ for $t \in (a, b)$, then f is constant.*

Proof First we derive a contradiction when $f''(t_0) < 0$. Assume that $f''(t_0) \leq -\delta < 0$. This implies that there exists $\alpha \in \mathbb{R}$ such that

$$f(t_0) \leq f(t) \leq f(t_0) + \alpha(t - t_0) + \frac{-\delta + \epsilon}{2}(t - t_0)^2.$$

This forces $\alpha = 0$ from which it follows that the support function must also have a minimum at t_0. This is a contradiction when $\epsilon < \delta$.

In general suppose that there exists $t_1 \geq t_0$ such that $f(t_1) > f(t_0)$, i.e., f is not constant on $[t_0, b]$. Consider the auxiliary function:

$$\varphi(t) = f(t) + \delta(1 - e^{t-t_0}).$$

When $t \leq t_0$

$$\varphi(t) = f(t) + \delta(1 - e^{t-t_0}) \geq f(t) \geq f(t_0)$$

and for sufficiently small δ it follows that

$$\varphi(t_1) > \varphi(t_0) = f(t_0).$$

So $\varphi(t)$ must have a minimum in $[t_0, t_1) \subset (a, b)$. However, as

$$\varphi''(t) \leq f''(t) - \delta e^{t-t_0} \leq -\delta e^{t-t_0} < 0$$

we have obtained a contradiction. A similar argument works when f is not constant on $[a, t_0]$ using

$$\varphi(t) = f(t) + \delta(1 - e^{t_0-t}).$$

□

Corollary 3.4 *If $f'' \leq 0$ on I, then f is concave. Moreover, if $f'' \leq 0$ or more generally $f'' \leq \lambda$ on I, then the one sided derivatives $f'^{\pm}(t_0) = \lim_{t \to t_0^{\pm}} \frac{f(t)-f(t_0)}{t-t_0}$ exist.*

Proof The minimum principle shows that if the graph of f intersects a straight line in two or more points, then it either coincides with the line on an interval $[a, b] \subset I$ or otherwise lies above the line. □

Remark 3.5 It is worth noting that the calculus of Newton as developed in Principia is rigorous when based on these principles of monotonicity and/or concavity. For example, it is very easy to prove that bounded monotone functions are integrable, a result that Newton commences with in Principia.

Exercise 3.6 With that in mind it is interesting to note that if $f : [0, \infty) \to [0, \infty)$, and $f^{(n)} \geq 0$ for all $n \in \mathbb{N}$, i.e., f and all its derivatives are increasing, then f is analytic. This can be shown using either the mean value or integral version for the error terms in Taylor polynomials.

The minimum principle leads us to a more general comparison for functions that are sub-solutions to second order linear differential equations.

Theorem 3.7 (Main Comparison Result) *Assume $f : [a, b] \to \mathbb{R}$ continuous, $f'' + kf \leq 0$, $f(a) = 0$, $f'^{+}(a) = 0$. It follows that $f(t) \leq 0$ for $t \leq b$ when $k \leq 0$ and $f(t) \leq 0$ for $t \leq \frac{\pi}{\sqrt{k}} + a$ when $k > 0$.*

Proof By scaling we assume that $k = 0, \pm 1$. When $k = 0$ we have $f'' \leq 0$, $f(a) = 0$, $f'(a) = 0$, so f is concave and monotone.

When $k = -1$ consider $h(t) = f(t) - \epsilon \sinh(t)$. We have $h'' - h \leq 0$, $h(a) = 0$, $h'(a) = -\epsilon < 0$. This shows that h is negative for small t. If there is a t_0 such that $h(t_0) > 0$, then h must have a negative minimum on $[a, t_0]$. Hence $h'' \leq h < 0$ at this minimum, which is a contradiction. Thus $h(t) \leq 0$ for all t. This implies that $f(t) \leq \epsilon \sinh(t)$ for all ϵ and we obtain the conclusion by letting $\epsilon \to 0$.

The case where $k = 1$ is trickier. For simplicity we translate along the t-axis so that $a = 0$. This time we need to consider the more complicated function

$$h(t) = \frac{f(t) - \epsilon \sin(t)}{\sin((1 + \delta)t)} = \frac{f_\epsilon(t)}{\zeta_\delta(t)}, \; h(0) = 0, \; h'(0) < 0.$$

This will evidently also have a negative minimum if it becomes positive at some later time. If the minimum is at t_0, then a direct calculation shows that:

$$\begin{aligned} h''(t_0) &\leq \frac{f_\epsilon''(t_0)}{\zeta_\delta(t_0)} - 2\frac{\zeta_\delta'(t_0)}{\zeta_\delta(t_0)} \frac{d}{dt}\left(\frac{f_\epsilon}{\zeta_\delta}\right)\Big|_{t=t_0} - \frac{f_\epsilon(t_0)}{\zeta_\delta(t_0)} \frac{\zeta_\delta''(t_0)}{\zeta_\delta(t_0)} \\ &\leq -\frac{f_\epsilon(t_0)}{\zeta_\delta(t_0)} + \frac{f_\epsilon(t_0)}{\zeta_\delta(t_0)}(1 + \delta)^2 \\ &= \left(2\delta + \delta^2\right) \frac{f_\epsilon(t_0)}{\zeta_\delta(t_0)} \\ &< 0, \end{aligned}$$

which is a contradiction. This argument works as long as $(1+\delta)t \leq \frac{\pi}{\sqrt{k}}$ and leads us to conclude that $f_\epsilon(t) \leq 0$. First letting $\epsilon \to 0$ shows that $f(t) \leq 0$ for $(1 + \delta)t \leq \frac{\pi}{\sqrt{k}}$. We can then finally let $\delta \to 0$ to obtain the claim. □

3.2 Calculus on Alexandrov Spaces

We start by showing how certain modified distance functions on the space forms, S^n_k, satisfy these types of second order linear differential equations.

$k = 0$ On $(\mathbb{R}^n, g_0)$ use $r(x) = |xp| = |x - p|$ and observe that $\frac{1}{2}r^2 = \frac{1}{2}|x - p|^2$ is smooth everywhere with $\mathrm{Hess}(\frac{1}{2}r^2) = g_0$.

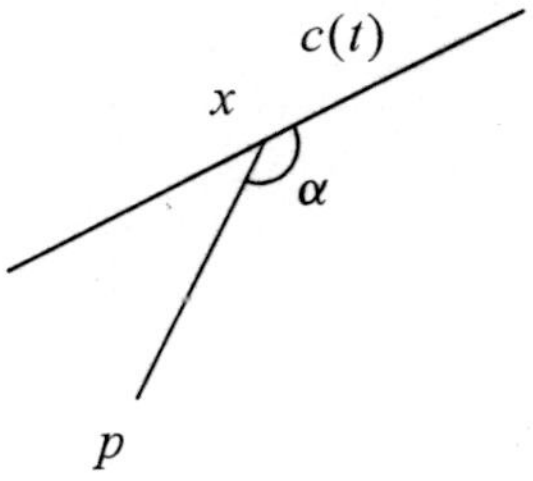

Now select a unit speed geodesic $c(t)$. The second derivative of $f(t) = \frac{1}{2}|c(t) - p|^2$ is 1, i.e., $f'' = 1$. Integration of $f'' = 1$ leads to the Euclidean law of cosines with $x = c(0)$:

$$\frac{1}{2}|c(t) - p|^2 = \frac{1}{2}t^2 + \frac{1}{2}|p - x|^2 + t|p - x|\cos(\pi - \alpha).$$

$k = 1$ This time consider the function $1 - \cos(|xp|) = \frac{1}{2}|xp|^2 + O(|xp|^3)$ which satisfies $\mathrm{Hess}(1 - \cos(|xp|)) = \cos(|xp|)g_{S^n}$. The motivation comes from the fact that in $\mathbb{R}^{n+1}$ we have $\cos|xp| = x \cdot p$. If we consider $f(x) = x \cdot p$ on $\mathbb{R}^{n+1}$, then $df(v) = v \cdot p$ and $\mathrm{Hess}\, f = 0$. Using x as the unit normal to S^n we can then calculate

$$\begin{aligned}
\mathrm{Hess}\, f|_{S^n}(X, Y) &= \left(\nabla^{S^n}_X df\right)(Y) \\
&= D_X(df(Y)) - df\left(\nabla^{S^n}_X Y\right) \\
&= \mathrm{Hess}\, f(X, Y) + df\left(x\left(x \cdot \nabla^{\mathbb{R}^{n+1}}_X Y\right)\right) \\
&= (x \cdot p)\, \mathrm{II}(X, Y) \\
&= -x \cdot p g_{S^n}(X, Y).
\end{aligned}$$

Evaluation on a geodesic leads to the equation $f'' + f = 1$, which in turn can be used to establish the spherical law of cosines.

$k = -1$ In analogy with the sphere we think of $H^n \subset \mathbb{R}^{n,1}$. The inner product in $\mathbb{R}^{n,1}$ is given by $v \cdot w = v^1w^1 + \cdots + v^nw^n - v^{n+1}w^{n+1}$ and $H^n = \{x \in \mathbb{R}^{n,1} \mid x \cdot x = -1, x^{n+1} > 0\}$. We can use $x \cdot p - 1 = \cosh(|xp|) - 1$

with $\text{Hess}(\cosh(|xp|) - 1) = \cosh(|xp|)g_{H^n}$. This time $f'' - f = 1$, and we obtain the hyperbolic law of cosines.

Remark 3.8 The second order equations $f_k''+kf_k = 1$ also hold on the cones $C_k(X)$ and suspensions $\Sigma_1 X$ when curv $X \geq 1$ and p the vertex of the cone.

In these examples the distance functions were all modified so that their Hessians became conformal to the metric. There is even a general structure theorem for Riemannian metrics with conformal Hessians (proofs can be found in [25]).

Theorem 3.9 (Brinkmann (Local) and Toshiro (Global)) *Let (M, g) be a smooth complete Riemannian manifold. If $\text{Hess}(f) = \lambda g$ for some function $\lambda : M \to \mathbb{R}$, then M is a warped product, i.e., $g = dr^2 + \phi(t)^2 h$.*

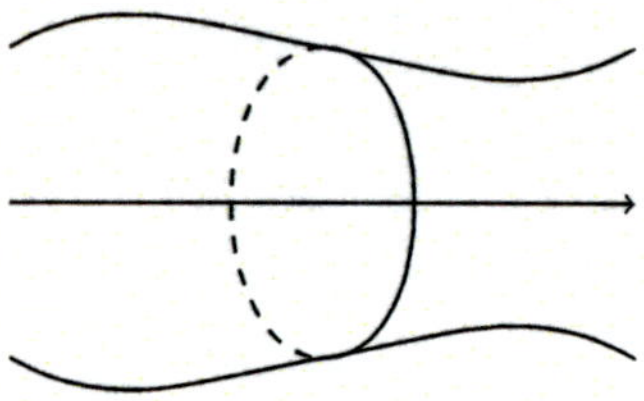

Definition 3.10 On an Alexandrov space $(X, |\cdot\cdot|)$ with curv $\geq k$ we say that $f : X \to \mathbb{R}$ satisfies $f''(p) \leq \lambda$ provided $(f \circ c)''(0) \leq \lambda$ for any unit speed geodesic $c : I \to X$, $c(0) = p$.

When curv $\geq k$ we can as on the space forms consider the modified distance functions:

$$f_k(x) = \begin{cases} \frac{1}{2}|xp|^2 & k = 0, \\ 1 - \cos(|xp|) & k = 1, \\ \cosh(|xp|) - 1 & k = -1. \end{cases}$$

It follows from distance comparison that these functions satisfy the differential inequalities,

$$f_k'' + kf_k \leq 1,$$

along unit speed geodesics.

As we just saw in S_k^n the same formula for the modified distance function $\bar{f}_k$ gives us the differential equations $\bar{f}_k''+k\bar{f}_k = 1$ along geodesics. When considering the differences $h(t) = f_k \circ c(t) - \bar{f}_k \circ \bar{c}(t)$ we evidently obtain the differential inequalities $h'' + kh \leq 0$ considered above. Here we can specify initial conditions by choosing the geodesics $c, \bar{c}$ appropriately. This leads among other things to a proof of Toponogov's theorem if we know that $f_k'' + kf_k \leq 1$ holds for all x, p.

Next we wish to define a differential $df_p : T_pX \to \mathbb{R}$ for semi-concave functions. For this to work easily we additionally assume f is Lipschitz: $|f(x) - f(y)| \leq L|xy|$. Since $T_pX = C_0(S_pX)$ it suffices to construct $df_p : S_pX \to \mathbb{R}$. We can define the differential on directions that come from geodesics directly by

$$df_p(\dot{c}(0)) = \frac{d}{dt}|_{t=0^+}(f \circ c).$$

This is well-defined as $f \circ c$ is semi-concave. To show that the differential is uniformly continuous on this subset of directions assume for simplicity that curv $\geq$ 0. From triangle comparison we have

$$|c_1(t)c_2(t)|^2 \leq 2t^2 - 2t^2\cos(\theta) = 2t^2(1 - \cos(\theta)) \leq t^2\theta^2.$$

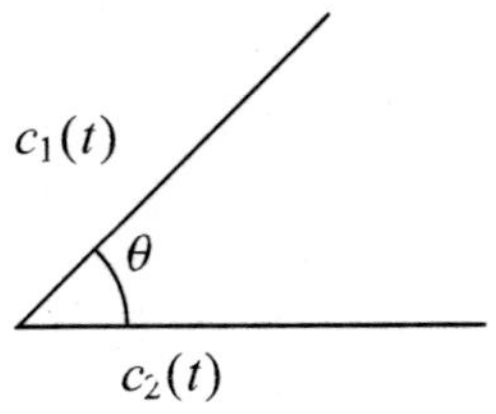

Thus we obtain

$$\frac{|f(c_1(t)) - f(c_2(t))|}{t} \leq \frac{L|c_1(t)c_2(t)|}{t} \leq L\theta = L\measuredangle(\dot{c}_1(0), \dot{c}_2(0)).$$

This shows that df_p is uniformly continuous on the geodesic directions and hence extends to a unique map on the space of directions.

This also allows us to define a gradient vector $\nabla f(p) \in T_pX$. The properties that determine this gradient are:

1. $\nabla f(p) = 0$ if $df_p : T_pX \to \mathbb{R}$ is ≤ 0.
2. $\nabla f(p) = df_p(v_0)v_0$ if $df_p : S_pX \to \mathbb{R}$ is maximal at v_0.

For this characterization to be meaningful we need v_0 to be unique. This will use that $T_pX = \lim_{\mu\to\infty}(X, \mu|\cdot\cdot|)$. The scaled function $\mu f : (X, \mu|\cdot\cdot|) \to \mathbb{R}$ is now $\frac{\lambda}{\mu}$-semi-concave and L-Lipschitz. By restricting to geodesics emanating from p we see that in fact $\lim_{\mu\to\infty} \mu f = df_p$. Thus $df_p : T_pX \to \mathbb{R}$ is 0-concave, in other words concave. This implies that $df_p : S_pX \to \mathbb{R}$ has a unique maximum on the set where it is positive.

It turns out that for distance functions these constructions are much simpler. Let $f(x) = |xC|$, where $C \subset X$ is a closed subset. The directional derivative can be calculated via the *first variation formula*:

$$df_p(v) = -\cos\theta_v, \ \theta_v = \min\left\{\angle\left(\overrightarrow{pq}, v\right) \mid q \in C\right\}.$$

To be clear, the minimum runs over all possibilities for minimal geodesics from p to C, even when C is a point.

This shows that $df_p \leq 0$ precisely when p is a *critical point* for f, i.e., for every $v \in S_pX$ there is exists $\overrightarrow{pq}$, $q \in C$, such that $\angle\left(\overrightarrow{pq}, v\right) \leq \frac{\pi}{2}$.

When, on the other hand, p is not critical, then $\left\{v \in S_pX \mid \theta_v \geq \frac{\pi}{2}\right\}$ is π-convex (see Proposition 5.2) with nonempty interior. Moreover, as we shall see in Sect. 4, the distance function to the boundary where $\theta_v = \frac{\pi}{2}$ is strictly concave on the interior. Consequently, there is a unique point $v_0 \in S_pX$ at maximal distance from the boundary. Since this direction maximizes θ_v we see that $\nabla f(p) = df(v_0)\, v_0$.

3.3 Special Curves

Aside from geodesics there are two types of very important special curves that appear in the subject of Alexandrov spaces: *gradient curves* and *quasi-geodesics*. The latter are extremely difficult to construct but they do make some proofs considerably simpler and more direct. However, often gradient curves can be used instead.

Definition 3.11 $c(t)$ is a quasi-geodesic if it is unit speed and all semi-concave f become semi-concave when restricted to c.

Theorem 3.12 *Given any $v \in S_pX$ there exists a not necessarily unique quasi-geodesic $c : [0, \infty) \to X$ such that $\dot{c}^+(0) = v$.*

The proof can be found in [27] and is quite long and intricate. From the definition of quasi-geodesics we see that when curv $\geq k$, then the differential inequalities,

$$f_k'' + kf_k \leq 1,$$

also hold along quasi-geodesics.

Gradient curves are by contrast not nearly as difficult to construct. Fix a semi-concave Lipschitz map $f : (X, |\cdot\cdot|) \to \mathbb{R}$ with $f'' \leq \lambda$. A gradient curve for f is a curve $c(t)$ such that

$$\dot{c}^+(t) = \nabla f(c(t)).$$

Such curves are forward unique. The gradient is not a continuous function but this ODE can still be solved by a construction similar to Euler's method (see [27]).

Here is a picture:

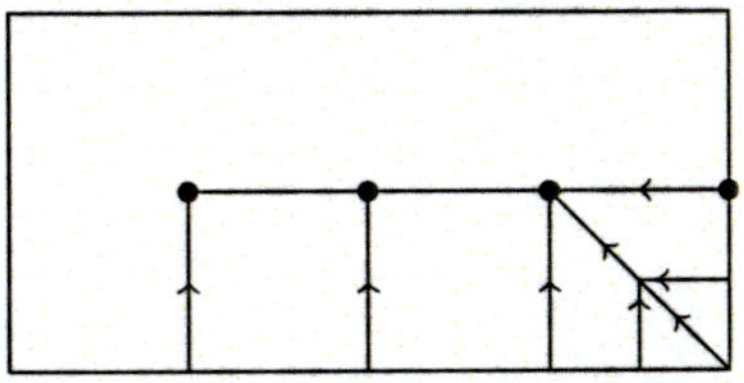

Here R is a rectangle and $f(x) = |x\partial R|$ on the inside. The gradient curves do bifurcate when going backwards but all move in the direction that increases the distance to the boundary the fastest. This is an example of the *Sharafutdinov retraction*.

Extremal sets in X are those sets which preserve gradient flows. In other words, $E \subset X$ is extremal provided any gradient curve $c(t)$ for any semi-concave Lipschitz map has the property that if $c(t_0) \in E$, then $c(t) \in E$ for all $t \geq t_0$. It is not hard to check that a point $p \in X$ is extremal when diam $S_pX \leq \frac{\pi}{2}$ as any function must have p as a critical point. More importantly, the boundary $\partial X \subset X$ is always an extremal set. One can also show that any unit speed curve in an extremal set that realizes the intrinsic distance between its endpoints, i.e., an intrinsic geodesic, is a quasi-geodesic in the ambient space. In particular, curves in ∂X that are intrinsically geodesics are quasi-geodesics in X (see [27]).

One concept that is crucial in Riemannian geometry is the exponential map: $\exp_p : T_pM \to M$. On Alexandrov spaces the domain of this map is restricted to the subset of the tangent cone that corresponds to geodesics emanating from p. Instead we have a gradient exponential map $\mathrm{gexp}_p : T_pX \to X$ with the property that $c(t) = \mathrm{gexp}_p(tv)$, $v \in S_pX$, is the gradient curve for $|xp| = r(x)$ that starts in the direction of v. Since minimal unit speed geodesics are automatically gradient curves we see that $\mathrm{gexp}_p \circ \log_p(x) = x$ for all $x \in X$.

When curv ≥ 0 the map $\mathrm{gexp}_p : T_pX \to X$ is 1-Lipschitz:

$$|\mathrm{gexp}_p(v)\mathrm{gexp}_p(w)| \leq |vw|.$$

This is better than the exponential map which only satisfies this on the region inside the cut-locus. In fact, the gradient exponential could become an excellent tool in traditional Riemannian geometry.

Another use of gradient curves allows us to conclude that if we have a distance function $f(x) = |xC|$, $C \subset X$ is compact, then $f^{-1}((a, \infty))$ deformation retracts to $f^{-1}([b, \infty))$, provided f has no critical points in $f^{-1}((a, b))$. However, as we shall see momentarily, a much stronger result holds.

3.4 The Fibration and Stability Theorems

A map $F : X \to \mathbb{R}^k$, where the coordinate functions F^i are λ-concave, is called *tight* provided

$$\measuredangle\left(\nabla F^i, \nabla F^j\right) > \frac{\pi}{2}, \text{ for all } i \neq j.$$

For such maps a point $p \in X$ is said to be *regular* if there exists $v \in S_pX$ such that also

$$\measuredangle\left(\nabla F^i, v\right) > \frac{\pi}{2}.$$

In case $T_pX = \mathbb{R}^n$ this guarantees that the gradients ∇F^i are linearly independent. Note that when $k = 1$ any λ-concave function is tight, and a point is regular when its gradient is nonzero.

Perel'man's Fibration Theorem states that if $F : U \subset X \to F(U) \subset \mathbb{R}^k$ is a proper tight map and without critical points, then it is a fibration. This clearly requires that $k \leq \dim X$. It is proven by reverse induction on k starting with $k = \dim X$, where the assertion is that the map becomes locally injective. This is a nice generalization of what happened with strainers, but it is not a complete replacement as it requires that we know X has finite (Hausdorff) dimension.

The local version of the Stability Theorem is even more subtle and is, in fact, based on the Fibration Theorem. For us the most important feature is that it shows that Alexandrov spaces are pointed locally homeomorphic to their tangent cones. In particular, as all compact 1-dimensional Alexandrov spaces are circles or closed intervals, it follows that all 2-dimensional Alexandrov spaces are 2-manifolds. This in turn tells us that the only compact 2-dimensional spaces with positive curvature are homeomorphic to spheres, discs, or real projective planes. To be clear, this uses that the space and its universal covering have a positive lower curvature bound, which in turn implies that it is compact (see also Sect. 4). From this we see that 3-dimensional Alexandrov spaces have the property that a point that is not a manifold point is an orbifold singularity which is modeled on a cone over the real projective plane. This can in turn be used to give a topological classification of compact 3-dimensional Alexandrov spaces with positive curvature (see [9]). When they have no boundary they are homeomorphic to space forms S^3/Γ or $\Sigma\mathbb{RP}^2$, and when they have boundary they are cones over S^2 or $\mathbb{RP}^2$. This helps to show that 4-dimensional Alexandrov spaces are orbifolds.

The global version of the Stability Theorem is the easiest of the results to state and the most difficult to prove. Given a compact n-dimensional Alexandrov space X, there exists an $\epsilon = \epsilon(X) > 0$, such if Y is a compact n-dimensional Alexandrov space with Gromov-Hausdorff distance, $d_{G-H}(X, Y) < \epsilon$, then X and Y are homeomorphic.

4 Rigidity Results

In this section we present the rigidity results that have formed the main focus of our recent research. Some shorter proofs are included in this section, while the more detailed and complicated proofs can be found in the next section.

We start with a more classical problem that gives the flavor of the proofs for the results mentioned below.

4.1 Bonnet-Myers' and Toponogov's Diameter/Radius Theorem

For a general metric space we say that $\operatorname{rad} X \leq r$ if there is $p \in X$ such that $X = \bar{B}(p, r)$.

Theorem 4.1 ([4]) *An Alexandrov space with* $\operatorname{curv} X \geq 1$ *has* $\operatorname{diam} \leq \pi$ *and* $\operatorname{diam} = \pi$ *implies that* $X = \Sigma_1(S_p X)$ *for some* $p \in X$. *Moreover, if* $\operatorname{rad} X = \pi$, *then* $X = S^n$ (1).

Proof To prove the first statement select $p^{\pm}$ such that $2\pi >> |p^+ p^-| \geq \pi$. For any x it follows that $|p^+ x| + |xp^-| \geq |p^+ p^-| \geq \pi$. The functions $f_{\pm}(x) = 1 - \cos(|xp^{\pm}|)$ satisfy $f_{\pm}'' + f_{\pm} \leq 1$. This shows that

$$(f_+ + f_-)'' \leq 2 - f_+ - f_- = \cos(|xp^+|) + \cos(|xp^-|) \leq 0,$$

when $|xp^+| + |xp^-| \geq \pi$ and $|xp^+| \leq \pi$, $|xp^-| \leq \pi$.

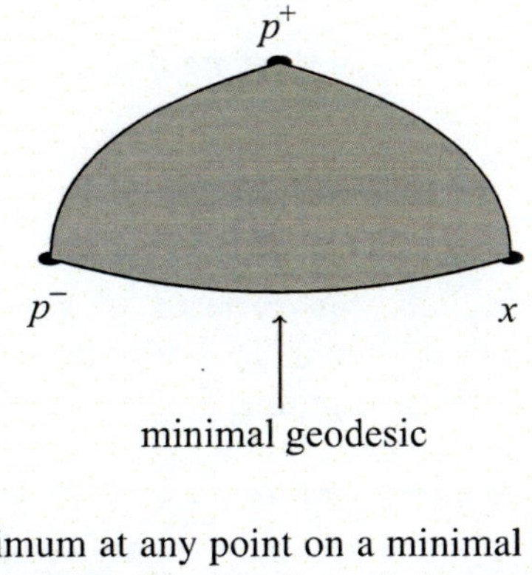

But $f_+ + f_-$ has a minimum at any point on a minimal geodesic from p^+ to p^-. Thus the minimum principle implies that $f_+ + f_-$ is constant. If $|p^+ p^-| > \pi$, then $f_+ + f_-$ can't be constant. When $|p^+ p^-| = \pi = \operatorname{diam}(X)$, then $|p^+ x| + |xp^-| = \pi$ for all $x \in X$. In particular, any x lies on a geodesic from p^+ to p^-. Thus

$$f_+'' \leq 1 - f_+ = \cos(|xp^+|) = \cos(|p^- x|) \leq -f_-'' = f_+''.$$

Showing that $f_{\pm}'' + f_{\pm} = 1$. To see that $X = \Sigma_1(S_pX)$ with p being either of $p^{\pm}$ we need to show that any direction at p is a direction for a geodesic between p^+ and p^-. Since any $x \in X - \{p^{\pm}\}$ lies on a geodesic from p^+ to p^- it follows that any geodesic direction at p corresponds to such a geodesic. Since any direction at p is a limit of geodesic directions such a direction is the limit of a sequence of geodesics from p^+ to p^-. We can now apply Arzela-Ascoli to finish the claim. This shows that $\log_p : X \to \Sigma_1 S_pX$ defined by $x \mapsto \left(|px|, \overrightarrow{px}\right)$ is well-defined, a homeomorphism, and distance nondecreasing. The fact that the equations $f_{\pm}''+f_{\pm} = 1$ hold along geodesics in X shows that they remain geodesics when mapped to $\Sigma_1 S_pX$. This implies that X and $\Sigma_1 S_pX$ are isometric.

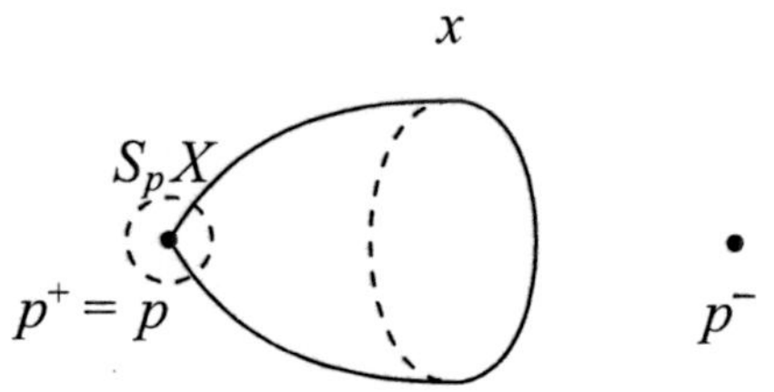

The second claim is established via induction on dimension by showing that also $\operatorname{rad} S_pX = \pi$. Let c be a geodesic with $\dot{c}(0) \in S_pX$. If we consider $q^+ = c(t)$, then there is a q^- at distance π from q^+. By angle comparison this forces $\angle\left(\overrightarrow{pq^+}, \overrightarrow{pq^-}\right) = \pi$ showing that each geodesic direction has a direction at maximal angle π from it. Since geodesic directions are dense we have proved the claim.

To get the induction started we simply need to note that in dimension one only $S^1(1)$ and $[0, \pi]$ have maximal diameter and the latter only has radius $\frac{\pi}{2}$. □

Remark 4.2 Recall that the diameter rigidity statement for Riemannian manifolds shows that $M = S^n(1)$ since the space of directions is also a unit sphere. For Alexandrov spaces we clearly need to invoke the stronger radius assumption. Volume comparison further shows that when the volume is maximal, $\operatorname{vol} S^n(1)$, then so is the radius.

The suspension rigidity can be used to gain information about spaces of directions. For example, if a minimal geodesic passes through a point p, then $S_pX = \Sigma_1 Z$ and $Z \subset S_pX$ can be identified with the directions that are perpendicular to the minimal geodesic. A special case occurs when X has boundary ∂X. The double $D(X)$ has the property that for any $p \in \partial X$, $S_pD(X) = D\left(S_pX\right)$. So if p is chosen to be closest to a point in the interior, then the corresponding minimal geodesic will in the double create a minimal geodesic through p. In this case we can identify $Z = S_p\partial X = \partial S_pX$ and $S_pX = \{v\} * S_p\partial X$, where v is the direction of the minimal geodesic.

There are some interesting topological counter parts to the above rigidity phenomena that also play a role in our results. The first result appeared in [24]. It and the subsequent result can be found in [27]

Theorem 4.3 (Perel'man) *If* $\operatorname{curv} X \geq 1$ *and* $\operatorname{diam} X > \frac{\pi}{2}$, *then* X *is homeomorphic to the suspension* $\Sigma S_p X$ *for a suitable* $p \in X$.

Here p is chosen so that $|pq| = \operatorname{diam} X$ for some $q \in X$ and the proof uses the Fibration theorem applied to the distance function from p. This function has the property that only p, q are critical points. This is easy to prove via angle comparison (see also proposition 4.2).

Theorem 4.4 (Petrunin) *If* $\operatorname{curv} X \geq 1$ *and* $\operatorname{rad} X > \frac{\pi}{2}$, *then* $\operatorname{rad} S_p X > \frac{\pi}{2}$ *for all* $p \in X$.

The proof of this result is similar to Proposition 5.1. The theorem leads to an elegant proof of the radius sphere theorem (see [11, 27]).

Theorem 4.5 (Grove and Petersen) *If* $\operatorname{curv} X \geq 1$ *and* $\operatorname{rad} X > \frac{\pi}{2}$, *then* X *is homeomorphic to a sphere.*

4.2 Controlling the Boundary

We are now ready to explain some of the rigidity results that people have considered more recently. Assume that X^n has dim $= n$, curv $\geq k$, and $\partial X \neq \emptyset$. The first obvious result to notice is that: if $\operatorname{rad}(X) \leq r$, then $\operatorname{vol}(X) \leq \operatorname{vol}(\bar{B}_k(r))$, where $B_k(r) \subset S_k^n$, and equality only occurs when $X = \bar{B}_k(r)$. The inequality for the volume is simply volume comparison while the rigidity statement can be proven like the maximal diameter/radius result above. Further details of the proofs of such rigidity statements will be given below.

There are two natural and considerably more challenging questions one can ask. The volume of the boundary is the $(n-1)$-dimensional Hausdorff measure with respect to the intrinsic metric.

1. If $\operatorname{rad}(X) \leq r$, is it true that $\operatorname{vol}(\partial X) \leq \operatorname{vol}(\partial \bar{B}_k(r))$, where $B_k(r) \subset S_k^n$?
2. What conditions on ∂X guarantee $\operatorname{rad} X \leq r$?

To address the first question one needs a 1-Lipschitz gradient exponential map $\operatorname{gexp}_p(k, \cdot) : C_k(S_p X) \to X$, where $C_k(S_p X)$ is the cone with radial curvature k. Here $C_0(S_p X) = T_p X$ and $C_1(S_p X) = \{o_p\} * S_p X$, i.e., just half of the spherical suspension. The restriction on the size of the spherical cone will not impose further assumptions as rad $\leq \frac{\pi}{2}$ when curv ≥ 1 and $\partial X \neq \emptyset$ (see also Lemma 5.3). For $k = 0$ this is the usual gradient exponential map, while for $k \neq 0$ the gradient curves need to be reparametrized so as to make the map distance nonincreasing.

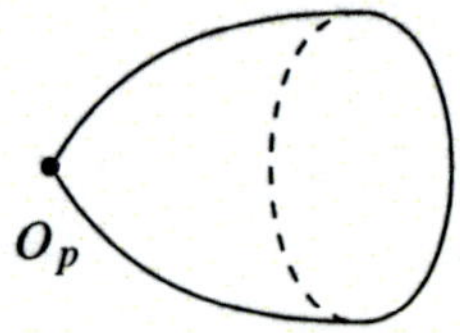

half of $\Sigma(S_p X)$

Petrunin in [27] establishes a positive answer to the first question as follows: First one shows that $\text{gexp}_p(k, \partial \bar{B}(o_p, r)) \supset \partial X$. In other words: any point on the boundary is hit by a gradient curve. An obvious choice is a minimal geodesic from p to the point on the boundary. This shows that $\text{vol}\, \partial X \leq \text{vol}\, \partial \bar{B}(o_p, r) \leq \text{vol}\, \partial \bar{B}_k(r)$.

The most obvious way of answering the second question involves quantifying the convexity of the boundary. Note that $\partial \bar{B}_k(r) \subset S_k^n$ is totally umbilic: $\Pi_\partial = \lambda_k(r) g_\partial$, where

$$\lambda_k(r) = \begin{cases} \frac{1}{r} & k = 0, \\ \cot(r) & k = 1, \\ \coth(r) & k = -1. \end{cases}$$

In $\bar{B}_k(r_0)$ consider $d_k(x) = \phi_k(|x \partial \bar{B}_k(r_0)|)$, where

$$\phi_k'' + k\phi_k = -\lambda_k(r_0) = -\lambda_0, \; \phi_k(0) = 0, \; \phi_k'(0) = 1.$$

Specifically,

$$\phi_k(t) = \begin{cases} t - \frac{\lambda_0}{2} t^2 & k = 0, \\ \sin(t) + \lambda_0 \cos(t) - \lambda_0 & k = 1, \\ \sinh(t) - \lambda_0 \cosh(t) + \lambda_0 & k = -1. \end{cases}$$

This modified distance function on $\bar{B}_k(r)$ satisfies the second order equation:

$$d_k'' + k d_k = -\lambda_0.$$

Definition 4.6 We say that ∂X is λ_0-convex if $d_k(x) = \phi_k(|x \partial X|)$ satisfies $d_k'' + k d_k \leq -\lambda_0$. We also say that ∂X is strictly convex if it is λ_0-convex for some $\lambda_0 > 0$.

This definition conforms with the condition that $\Pi_\partial \geq \lambda_0 g_0$ for a Riemannian manifold with boundary ∂ (for a proof when $\lambda_0 = 0$ and $k = 0$ see [25]). In the Riemannian case it is also easy to see that this implies $\text{rad}(X) \leq r_0$ when $\lambda_k(r_0) = \lambda_0$. With this in place we can now look at some rigidity results.

Remark 4.7 Note that ϕ_k is chosen so that if c is a unit speed geodesic with $c(0) = p \in \partial X$ and $\dot{c}(0) \perp S_p\partial X$, then $(d_k \circ c)(t) = \phi_k(t)$.

Theorem 4.8 *Let X be an Alexandrov space with curvature $\geq k$ and boundary that is λ_0-convex, where $\lambda_0^2 > \max\{-k, 0\}$. The radius satisfies $\operatorname{rad} X \leq r_0$ with equality only when X is isometric to the cone $C_k(S_sX)(r_0) = \bar{B}(o_s, r_0) \subset C_k(S_sX)$ of radius r_0 for some $s \in X$.*

This was proved for $k \geq 0$ in [7] and with a different proof and in complete generality in [12].

Proof We start by showing that no point in X can have distance $> r_0$ to the boundary. Let $c : [0, L] \to X$ be a minimal unit speed geodesic from ∂X to $c(L)$. The explicit relationship between r_0 and λ_0 and the formula for $d_k \circ c$ show that $(d_k \circ c)'(L) < 0$ provided $L > r_0$, e.g., when $k = -1$ we have $(d_{-1} \circ c)(t) = \sinh(t) - \lambda_0 \cosh(t) + \lambda_0$ and $\lambda_0 = \coth(r_0)$, which yields

$$(d_{-1} \circ c)'(t) = \cosh(t) - \coth(r_0)\sinh(t) = \frac{\sinh(r_0 - t)}{\sinh(r_0)}.$$

Having $(d_k \circ c)'(L) < 0$ shows that $(d_k \circ c)(t)$ reaches a maximum before L and thus that c cannot be a minimal geodesic from $c(L)$ to ∂X. This in turn implies that d_k is strictly concave. For our modified distance functions this is obvious when $k \geq 0$ as we have $d_k'' \leq -kd_k - \lambda_0$. When $k = -1$ the Hessian is also negative away from the boundary as:

$$d_k'' \leq d_k - \lambda_0 = \sinh r - \coth r_0 \cosh r = -\frac{\cosh(r - r_0)}{\sinh r_0}.$$

This shows that X has a unique point soul $s \in X$ at maximal distance $r_1 \leq r_0$ from the boundary. Consider a quasi-geodesic $c : [0, b] \to X$ with $c(0) = s$ and the function $d_k(t) = d_k \circ c$. This function satisfies

$$d_k'' + kd_k \leq -\lambda_0,\ d_k(0) = \phi(r_1),\ d_k'(0) \leq 0.$$

And by comparison $d_k \leq \bar{d}_k$, where

$$\bar{d}_k{}'' + k\bar{d}_k = -\lambda_0,\ \bar{d}_k(0) = \phi(r_1),\ \bar{d}_k{}'(0) = 0.$$

The explicit form of $\bar{d}_k$ is as follows

$$\bar{d}_k = \begin{cases} -\cot r_0 + (\sin r_1 + \cot r_0 \cos r_1)\cos t, & k = 1, \\ r_1 - \frac{r_1^2}{2r_0} - \frac{t^2}{2r_0}, & k = 0, \\ \coth r_0 - (\sinh r_1 - \coth r_0 \cosh r_1)\cosh t, & k = -1. \end{cases}$$

These expressions imply that $\bar{d}_k(r_0) \leq 0$, and that $\bar{d}_k(r_0) = 0$ only occurs when $r_1 = r_0$. This shows that $b \leq r_0$ and consequently that the radius is $\leq r_0$. Moreover, equality can only happen when $r_1 = r_0$, i.e., the boundary is at constant distance from the soul.

The remainder of the proof can be completed along the same lines as the diameter rigidity statement in Theorem 4.1. If we assume that $r_1 = r_0$, then this shows more generally that any quasi-geodesic emanating from the soul must hit the boundary precisely when $t = b = r_0$. Since this agrees with the distance to any point on the boundary any such quasi-geodesic is a minimal geodesic. In particular, any point in X lies on a minimal geodesic from the soul to the boundary and $f(x) + r(x) = |xs| + |x\partial X| = r_0$ for all $x \in X$. This shows that $f' = -r'$ and $r'' = -f''$ along any geodesic in X. This implies that $d_k'' + kd_k = -\lambda_0$ and the modified distance to the soul, f_k, satisfies $f_k'' + kf_k = 1$. This shows that $\log_p : X \to C_k(S_s X)(r_0)$ is well defined and an isometry. One can similarly see that the gradient exponential map $\mathrm{gexp}_s(k; \cdot) : C_k(S_s X)(r_0) \to X$ becomes an isometry, as the gradient flow for f_k from the soul is a flow along minimal geodesics and the distance between points on these radial geodesics is governed by the equation $f_k'' + kf_k = 1$. □

There is a more geometric definition of convexity that is more local in nature and is independent of the lower curvature bound.

Definition 4.9 An Alexandrov space X has λ_0-*convex* boundary, where $\lambda_0 > 0$, provided: For each $x \in \mathrm{int}X$ and $p \in \partial X$ with $|xp| = |x\partial X|$ we have

$$|pq| \cos\left(\angle\left(\overrightarrow{px}, \overrightarrow{pq}\right)\right) - \frac{\lambda_0}{2}|pq|^2 \geq o\left(|pq|^2\right)$$

for all $q \in \partial X$ sufficiently near p. The law of cosines shows that $\bar{B}_k(r_0)$ has λ_0-convex boundary in this sense.

It is not hard to see that the above definition of λ_0-convex boundary implies this more local version. The converse is also true, but considerably more involved (see [1]).

4.3 *The Positive Mass Conjectures*

The various versions of the positive mass conjectures for manifolds with boundary offer a further motivation for the above questions. The first version was formulated and proved by Miao in [22].

Conjecture 4.10 Consider a Riemannian manifold (M^n, g). If scal ≥ 0, $\partial M = S^{n-1}(1)$, and $\Pi_{\partial M} \geq g_{\partial B(0,1)}$; then M is a unit ball in Euclidean space.

Min-Oo in [23] established the hyperbolic equivalent and also conjectured a version for positive curvature.

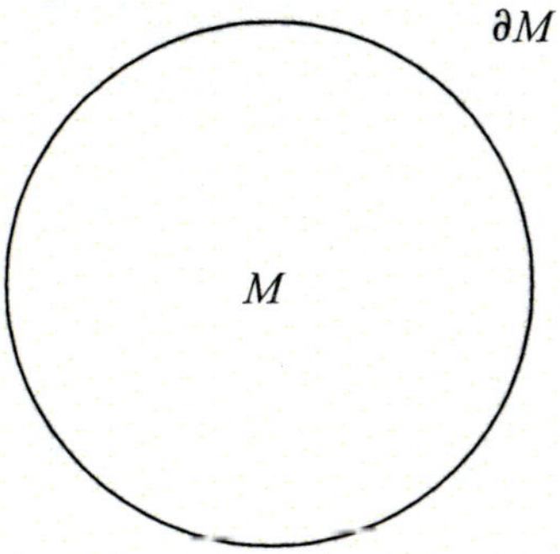

Conjecture 4.11 (Min-Oo [23]) If (M^n, g) **satisfies** $\mathrm{scal} \geq n(n-1)$, $\partial M = S^{n-1}(1)$, and $\Pi_{\partial M} \geq 0$; then M is a hemisphere.

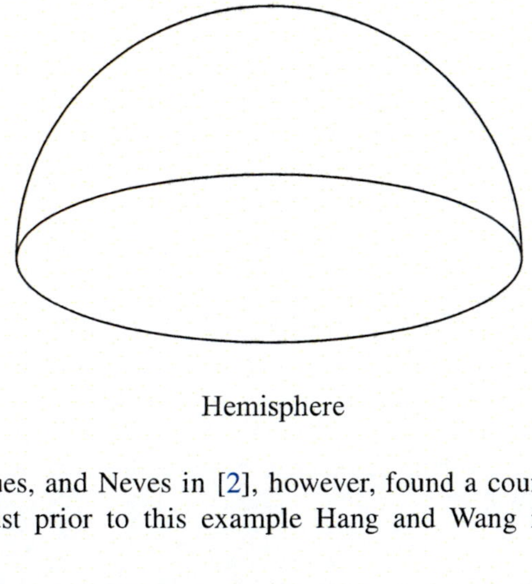

Hemisphere

Brendle, Marques, and Neves in [2], however, found a counterexample to this conjecture. But just prior to this example Hang and Wang in [14] proved the following version.

Theorem 4.12 *If the scalar curvature assumption is replaced with the condition:* $\mathrm{Ric} \geq n-1$, *then the conclusion of Min-Oo's conjecture holds.*

It is worth pointing out that this theorem is indeed extremely sensitive to the condition that the boundary be smooth. Even with the much stronger condition that $\sec \geq 1$ the so-called *Alexandrov lens*:

$$S^{n-2} * [-\alpha, \alpha] = \underbrace{\Sigma_1 \ldots \Sigma_1}_{n-1\text{-times}}([-\alpha, \alpha]), \alpha < \frac{\pi}{2}$$

is a counterexample. Here the boundary is convex but not smoothly embedded in the whole space. The two-dimensional version is a biangle or bigon. Note, however, that the boundary is intrinsically the unit sphere:

$$\partial \left(S^{n-2}(1) * [-\alpha, \alpha] \right) = S^{n-2}(1) * \{-\alpha, \alpha\} = S^{n-1}(1).$$

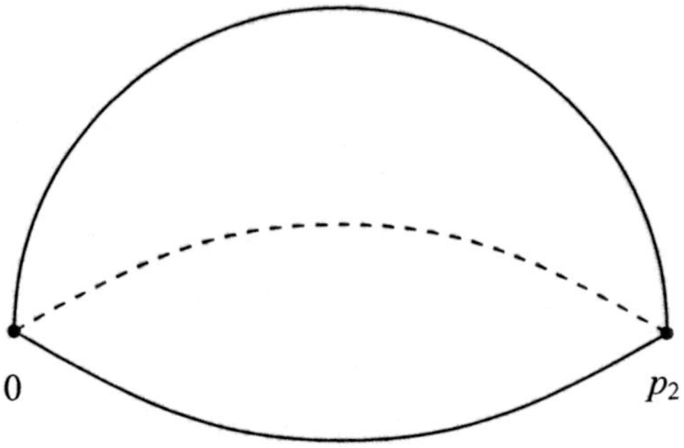

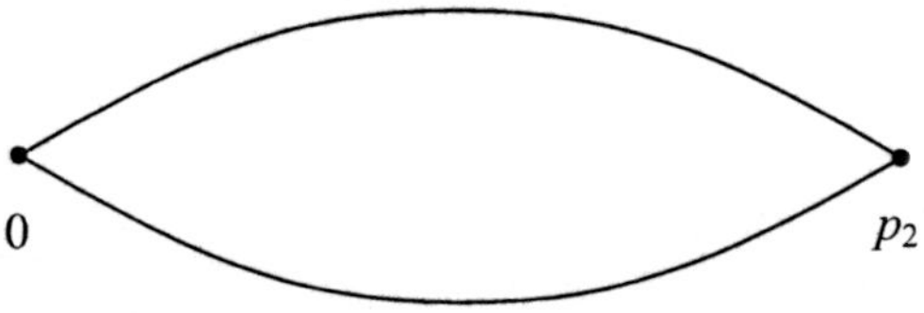

In general, this Alexandrov lens is best visualized as the intersection of two hemispheres that form an angle 2α. Note that the boundary is only convex, not strictly convex in the sense studied above.

With all of this in mind it would be interesting to determine which Riemannian n-manifolds with $\mathrm{Ric} \geq n-1$ and convex boundary have the property that $\mathrm{vol}\,(\partial M) = \mathrm{vol}\, S^{n-1}(1)$, both in the case of smooth boundary and more generally when it is merely metrically convex.

In case of lower bounds on curvature the complete answer is known and explained below.

4.4 *Lytchak's Problem*

We assume in the remainder of this section that X is an n-dimensional Alexandrov space with curv ≥ 1 and $\partial X \neq \emptyset$. Lytchak in a personal communication with Petrunin asked the following questions:

1. Is $\mathrm{vol}(\partial X) \leq \mathrm{vol}(S^{n-1}(1))$?
2. What happens when $\mathrm{vol}(\partial X) = \mathrm{vol}(S^{n-1}(1))$?

Petrunin addressed the first question as we explained above, but left the second question open as Lytchak's Problem. We just saw that there are indeed examples satisfying the conditions in the second question which are not cones or hemispheres. Thus this question is more subtle than the analogous question for spaces with strictly convex boundary as discussed in Theorem 4.8. Nevertheless, one can show (see [12]).

Theorem 4.13 (Grove, Petersen) *If an Alexandrov space X^n has* curv ≥ 1 *and* vol(∂X) = vol$(S^{n-1}(1))$, *then X is isometric to* $S^{n-2}(1) * [-\alpha, \alpha]$ *for some* $\alpha \leq \frac{\pi}{2}$.

With the weaker assumptions that curv ≥ 1, rad $= \frac{\pi}{2}$, and $\partial X \neq \emptyset$, there is not too much hope of a complete answer, again unlike the situations where the boundary is strictly concave. Below we give several examples that feature the complexities involved. We also offer several more general rigidity results, which indicate that rigidity might depend on the boundary looking either geometrically or topologically like a sphere.

To explain the examples we use a few facts that are established in the next section. Two sets $A, B \subset X$ are said to be *dual* to each other if $B = \{x \in X \mid |xA| \geq \frac{\pi}{2}\}$ and $A = \{x \in X \mid |xB| \geq \frac{\pi}{2}\}$. By standard comparison, both A and B are π-convex (proposition 4.2). In case X has boundary it will have a unique soul $s \in X$ at maximal distance from the boundary. We set $P = \{x \in X \mid |xs| \geq \frac{\pi}{2}\}$ and $Q = \{x \in X \mid |xP| \geq \frac{\pi}{2}\}$ and refer to them as the preliminary *edge* and *spine* of X. In case $\partial P = \emptyset$ we also use the notation $E = P$ and $S = Q$. Otherwise, we will show that there is an edge $E \subset \partial P$ and spine $S \supset Q$, where $\partial E = \emptyset$ and E and S are dual to each other. Moreover, the distance functions to E and to S have no critical points in $X - (E \cup S)$.

4.5 Examples

To check that a space has rad $= \frac{\pi}{2}$ we use the following proposition.

Proposition 4.14 *Assume that $A, B \subset X$ are convex subsets such that*

$$|ab| = \frac{\pi}{2} \text{ and } |Ax| + |xB| = \frac{\pi}{2}$$

for all $a \in A$, $b \in B$, and $x \in X$. If rad $B \leq \frac{\pi}{2}$ *and* rad $A \geq \frac{\pi}{2}$, *then* rad $X = \frac{\pi}{2}$.

Note that any spherical join $A * B$ with rad $B \leq \frac{\pi}{2}$ and rad $A \geq \frac{\pi}{2}$ satisfies the conditions of the proposition. Below we give examples which are quotients of such joins but not themselves spherical joins.

Example 4.15 The most basic construction is a spherical join $S * E$, where S has curv ≥ 1, $\partial S \neq \emptyset$, and rad $S < \frac{\pi}{2}$ and E has curv ≥ 1, $\partial E = \emptyset$, and rad $E \geq \frac{\pi}{2}$. To see how we might obtain such a decomposition consider a spherical join $[0, \pi] * [0, \pi]$, where both factors have boundary and radius $\frac{\pi}{2}$. This space has radius $\frac{\pi}{2}$ and can be rewritten as follows:

$$\begin{aligned}[0,\pi]*[0,\pi] &= [0,\pi]*\left(\left\{\frac{\pi}{2}\right\}*\{0,\pi\}\right)\\ &= \left([0,\pi]*\left\{\frac{\pi}{2}\right\}\right)*\{0,\pi\}\\ &= \left(\left[0,\frac{\pi}{2}\right]*\{0,\pi\}\right)*\{0,\pi\}\\ &= \left[0,\frac{\pi}{2}\right]*(\{0,\pi\}*\{0,\pi\})\\ &= \left[0,\frac{\pi}{2}\right]*S^1(1).\end{aligned}$$

Similarly, we have

$$\begin{aligned}[0,\alpha]*[0,\pi] &= \left([0,\alpha]*\left\{\frac{\pi}{2}\right\}\right)*\{0,\pi\}\\ &= \Sigma_1\left([0,\alpha]*\left\{\frac{\pi}{2}\right\}\right).\end{aligned}$$

Example 4.16 (Projective Lenses) Consider $E = \{0,\pi\}$ and $S = [-\alpha,\alpha]$, $\alpha < \frac{\pi}{2}$. Let $G = \mathbb{Z}_2$ be the natural reflection on both spaces. Note that $S*E = \Sigma_1[-\alpha,\alpha] = L^2_{2\alpha}$ looks topologically like a hemisphere. The action fixes $0 \in [-\alpha,\alpha]$ and acts like the antipodal map on $\partial(S*E) = \Sigma_1\{-\alpha,\alpha\} = S^1(1)$. The quotient looks topologically like a cone with vertex 0. The boundary has one point at distance $\frac{\pi}{2}$ from 0. The issue is that $\mathrm{rad}(E/G) < \frac{\pi}{2}$ and $\mathrm{rad}(X/G) < \frac{\pi}{2}$.

More generally one can consider the $\mathbb{Z}_2$ quotient of the Alexandrov lens $L^n_{2\alpha} = S^{n-2}(1)*[-\alpha,\alpha]$ (cf. also [8]). When $n > 2$, this space will have $\mathrm{rad} = \frac{\pi}{2}$ and boundary isometric to $\mathbb{RP}^{n-1}$.

Next we offer a more abstract version of the previous example that leads to more complicated examples.

Example 4.17 Assume we have an example $X = S*E$ as above and a compact group G that acts isometrically and effectively on S and E. This action is naturally extended to $S*E$ in such a way that it preserves the slices $S\times\{t\}\times E$ at constant distance from S and E. On these slices it is the natural action by G on $S\times E$. This leads to a new Alexandrov space X/G. Note that G preserves ∂X and ∂S and consequently also the common soul of both spaces. However, in the quotient, this "soul" need not be the soul of S/G or X/G, as $(\partial X)/G$ is not necessarily the boundary of the quotient. But it is true that the "soul" is at maximal distance from $(\partial X)/G$. Thus, $E/G \subset (\partial X)/G$ is at maximal distance $\frac{\pi}{2}$ from the "soul" and E/G and S/G are dual sets in X/G. It follows from the proposition that $\mathrm{rad}\,X/G = \frac{\pi}{2}$ provided $\mathrm{rad}\,E/G \geq \frac{\pi}{2}$. Topologically, $S*E$ is a cone over $\partial X = (\partial S)*E$, where the action fixes the soul and preserves the boundary. The quotient is likewise a topological cone over $(\partial X)/G$ where the "soul" is the vertex. This helps us identify the boundary of the quotient as $\partial(X/G) = ((\partial X)/G) \cup C(\partial((\partial X)/G))$, where $C(\partial((\partial X)/G))$ is topologically the cone over $(\partial X)/G$.

Example 4.18 (Preliminary Edge with Nonempty Boundary) Consider $Z = (\Sigma_1[-\alpha,\alpha])/\mathbb{Z}_2$ **and define** $X = S^1(r) * Z$, where $r \in \left[\frac{1}{2}, 1\right]$. The soul of X is the soul of Z and $P = C_1 S^1(r)$. This is an example where the preliminary edge P has nonempty boundary.

Further examples that indicate the complexities in trying to classify spaces with maximal radius can be obtained as follows:

Example 4.19 (Higher Dimensional Spines) Select $E = S^1(1)$ and $S = B(p,r) \subset S^2(1)$, where $r < \frac{\pi}{2}$. Let $G = \mathbb{Z}_2$ be a rotation by π on both S and E. This gives a 4-dimensional example where the boundary is homeomorphic to $\mathbb{RP}^3$. The same can be done with $E = S^n(1)$ and $S = B(p,r) \subset S^m(1)$, and $G = \mathbb{Z}_2$ the antipodal map on both S and E, giving an $n+m+1$-dimensional example with boundary homeomorphic to $\mathbb{RP}^{n+m}$.

Example 4.20 (Spines with Soul on Boundary) Select $E = S^1(1)$ and $S = B(p,r) \subset S^2(1)$, where $r < \frac{\pi}{2}$. Let $G = \mathbb{Z}_2$ be a rotation by π on E and a **reflection on** S. This gives a 4-dimensional example where the boundary of $S/\mathbb{Z}_2$ is connected and contains the soul of the quotient. The boundary is homeomorphic to a suspension $\Sigma\mathbb{RP}^2$. As in the previous example we can choose $E = S^n(1)$ and $S = B(p,r) \subset S^m(1)$, with $G = \mathbb{Z}_2$ action on E as the antipodal map and on S **as a reflection (or any other isometric involution). The resulting example is** $(n+m+1)$-dimensional with boundary homeomorphic to $\Sigma^{m-1}\mathbb{RP}^{n+1}$.

In both of these examples the key is that S is an Alexandrov space with curvature at least 1, non-empty boundary, and with an isometric involution.

In Corollary 5.19 we will show that above examples exhaust all the possibilities in dimensions ≤ 4, while the next example shows that one can have more complex behavior in dimensions ≥ 5.

Example 4.21 (Dual Pairs of Nonmaximal Dimension) Select $E = S^3(1)$ and $S = B(p,r) \subset S^2(1)$, where $r < \frac{\pi}{2}$. Let $G = S^1$ be the Hopf action on E and rotation around p on S. This gives a 5-dimensional example that is not **a finite quotient of a spherical join and with boundary homeomorphic to** $\mathbb{CP}^2$. Similar higher dimensional examples can be obtained with boundaries that are homeomorphic to complex and quaternionic projective spaces.

4.6 Generalizations of Lytchak's Problem

There are several results about spaces with curv ≥ 1, rad $= \frac{\pi}{2}$, and $\partial X \neq \emptyset$ that are more general than Lytchak's problem, but as mentioned, it is not yet clear what type of rigidity to expect in complete generality. The first two address what happens when the space of directions at the soul is geometrically spherical. These generalizations lead to a complete classification in dimensions up to four. All of the following results can be found in [12]

Theorem 4.22 (Grove and Petersen) *Let X be an n-dimensional Alexandrov space with* curv ≥ 1 *and $\partial X \neq \emptyset$. If* rad $X = \frac{\pi}{2}$ *and the soul of X is a regular point, then $E = S^k(1)$ and X is isometric to $S^k(1) * S$.*

Both this theorem and Theorem 4.13 can be proved using the following more general result. In Sect. 5 we will give a complete proof of this theorem.

Theorem 4.23 (Grove and Petersen) *Let X^n be an Alexandrov space with* curv ≥ 1, $\partial X \neq \emptyset$, *and* rad $X = \frac{\pi}{2}$. *If* rad $S_s X > \frac{\pi}{2}$, *then $X = E * S$, where* rad $E > \frac{\pi}{2}$ *and* rad $S_s S > \frac{\pi}{2}$.

Proofs of Theorems 4.13 and 4.22 Note first that when vol ∂X is maximal, then also vol $S_s X$ is maximal. This implies that $S_s X$ has maximal radius π and therefore that it is isometric to a unit sphere. Thus for both theorems we obtain a decomposition $X = E * S$, where the soul $s \in S$. This in turn implies that $S_s X = E * S_s S$. As $S_s X$ is a unit sphere both of E and $S_s S$ must also be unit spheres. This proves Theorem 4.22.

To prove Lytchak's problem we note that $\partial X = S^k(1) * \partial S$. Here ∂X cannot have maximal volume unless also ∂S has maximal volume. We can then use induction to finish the proof. □

Theorem 4.24 (Grove and Petersen) *Let X be an Alexandrov space with* curv ≥ 1, rad $X = \frac{\pi}{2}$, *and $\partial X \neq \emptyset$. When* dim $X \leq 4$, *then X is either isometric to a join or a $\mathbb{Z}_2$ quotient of a join, where $\mathbb{Z}_2$ acts effectively on both factors of the join.*

In case the boundary is known to be a Riemannian manifold we can also complement (if not complete) the main theorem in [8].

Theorem 4.25 (Grove and Petersen) *Let X be an n-dimensional Alexandrov space with* curv ≥ 1 *and boundary $\partial X = M^{n-1}$ that is intrinsically a Riemannian manifold with* sec $M \geq 1$. *If* rad $X = \frac{\pi}{2}$ *and M is topologically a sphere, then $X = S^{n-2}(1) * [-\alpha, \alpha]$.*

Proof In this case the edge E is a smooth totally geodesic submanifold of M with rad $\geq \frac{\pi}{2}$. From proposition 4.2 it follows that $M \cap S$ is a π-convex subset of M and dual to E in the sense of [10]. The arguments in [10] show that $M \cap S$ is a smooth totally geodesic submanifold of M without boundary. Since M is topologically a sphere with nontrivial dual submanifolds it follows again from [10] that all points in $M - (E \cup (S \cap M))$ lie on a unique minimal geodesic of length $\frac{\pi}{2}$ from E to

$S \cap M$. Further, for each $x \in S \cap M$ the corresponding map from the normal sphere to $S \cap M$ at x to E is a Riemannian submersion (and likewise for points in E). The classification of Riemannian submersions from spheres (cf. [10, 30]) and the fact that M is a topological sphere implies that $M = S^{n-1}(1)$. We can then use Theorem 4.13 to finish the proof. □

4.7 The Boundary Conjecture

The above results are focused on the space of directions at the soul, not just because that is more general than considering the boundary, but also because the general boundary conjecture has not yet been resolved.

Conjecture 4.26 (Boundary Conjecture) Is the boundary of an Alexandrov space with curv $\geq k$ also an Alexandrov space with curv $\geq k$?

This conjecture is known to be true for Riemannian manifolds with convex but not necessarily smooth boundary. Note that this includes convex sets in Euclidean space and so the boundary can actually have a very complicated and dense set of singularities. More recently the conjecture was settled for certain leaf spaces. In this case the boundary actually looks nicer and the proof depends on knowing that Lytchak's problem can be solved for such spaces (see [13] for the proof and further details on the boundary conjecture). As such, it formed the motivation for the above results in the more general class of Alexandrov spaces.

The first result in this direction was established in [13] and the next in [12].

Theorem 4.27 (Grove, Moreno, and Petersen) *The boundary conjecture is true for a leaf space X that comes from a submetry $f : M \to X$ where M is a Riemannian manifold and all preimages $f^{-1}(q) \subset M$ are properly embedded submanifolds.*

Using Theorem 4.25 this has now also been extended to a slightly more general case.

Theorem 4.28 (Grove and Petersen) *The boundary conjecture is true for a leaf space X that comes from a submetry $f : M \to X$ where M is a Riemannian manifold.*

5 The Weak Inner Regularity Theorem

In the final section we only consider Alexandrov spaces X with curvature ≥ 1. The purpose of this section is to prove Theorem 4.23. This requires numerous other constructions and results about various dual decompositions and their properties.

Many of these also appear virtually verbatim in the main reference [12]. We do deviate a little in a few cases so as to hopefully create a more complete and streamlined development. The first subsection is fairly self contained while the second relies on several results from the literature that we have not mentioned so far.

5.1 Basic Structure

Proposition 5.1 *If* $\operatorname{rad} X \geq \frac{\pi}{2}$ *and* $\operatorname{curv} X \geq 1$, *then for any* $p \in X$ *either* $\operatorname{rad} S_p X \geq \frac{\pi}{2}$ *or* $X = \Sigma_1 S_p X$.

Proof Assume $\operatorname{rad} S_p X < \frac{\pi}{2}$ and select a quasi-geodesic $c : [0, \pi] \to X$ such that $c(0) = p$ and $\angle(\dot{c}(0), w) < \frac{\pi}{2}$ for all $w \in S_p X$. We claim that if $q = c\left(\frac{\pi}{2}\right)$, then $|xq| < \frac{\pi}{2}$ unless $|px| = 0, \pi$. We have to use comparison along the quasi-geodesic as seen from x. This uses the modified distance $f_1(z) = 1 - \cos|zx|$ which satisfies $f_1'' + f_1 \leq 1$ along c and leads to

$$\begin{aligned}\cos|xq| &\geq \cos|xp| \cos\frac{\pi}{2} + \sin|xp| \sin\frac{\pi}{2} \cos\angle\left(\dot{c}(0), \overrightarrow{px}\right)\\ &= \sin|xp| \cos\angle\left(\dot{c}(0), \overrightarrow{px}\right).\end{aligned}$$

Since $\angle\left(\dot{c}(0), \overrightarrow{px}\right) < \frac{\pi}{2}$, the latter term is positive unless $\sin|xp| = 0$ and the result follows. In case $|px| = \pi$ we are finished. So if no such x exists then p is the one and only point at distance $\frac{\pi}{2}$ from q. This shows that $X - B(p, t) \subset \bar{B}\left(c(t), \frac{\pi}{2} - \delta(t)\right)$ for $t < \frac{\pi}{2}$ and t near $\frac{\pi}{2}$. As $c : \left[0, \frac{\pi}{2}\right] \to X$ is now a geodesic and $\angle(\dot{c}(0), w) < \frac{\pi}{2}$, for all $w \in S_p X$, it follows from Toponogov comparison that $\bar{B}(p, t) \subset \bar{B}\left(c(t), \frac{\pi}{2} - \epsilon(t)\right)$ for $t < \frac{\pi}{2}$. This shows that $X \subset B\left(c(t), \frac{\pi}{2}\right)$ for some $t < \frac{\pi}{2}$ and contradicts that $\operatorname{rad} X \geq \frac{\pi}{2}$. □

Proposition 5.2 *Let* X *be an Alexandrov space with* $\operatorname{curv} \geq 1$. *If* $C \subset X$ *is compact, then the dual set* $C^{\geq\frac{\pi}{2}} = \left\{x \in X \mid |xC| \geq \frac{\pi}{2}\right\}$ *is* π*-convex as is* $\partial X \cap C^{\geq\frac{\pi}{2}}$ *with respect to intrinsic geodesics in* ∂X. *Moreover, all critical points for* $x \mapsto |xC|$ *lie in* $C^{\geq\frac{\pi}{2}}$ *or the double dual* $\left(C^{\geq\frac{\pi}{2}}\right)^{\geq\frac{\pi}{2}}$.

Proof We don't need to consider the case where $C^{\geq\frac{\pi}{2}} = \emptyset$. Assume $x, y \in C^{\geq\frac{\pi}{2}}$ are joined by a geodesic c of length $< \pi$ and fix $p \in C$. By distance comparison $|p\,c(t)| \geq |\bar{p}\,\bar{c}(t)|$ where, $\bar{c}$ is a geodesic from $\bar{x}, \bar{y} \in S^2(1)$ of length $L(\bar{c}) = L(c)$ and $|px| = |\bar{p}\bar{x}|$, $|py| = |\bar{p}\bar{y}|$. However, in $S^2(1)$ any geodesic of length $< \pi$ between two points in the hemisphere of points at distance $\geq \frac{\pi}{2}$ from $\bar{p}$ must lie in that hemisphere, i.e., hemispheres are π-convex. This argument can repeated verbatim when c is merely a quasi-geodesic and thus, in particular, for intrinsic geodesics in ∂X.

To check the last statement select $p \in X - \left(C^{\geq \frac{\pi}{2}} \cup \left(C^{\geq \frac{\pi}{2}}\right)^{\geq \frac{\pi}{2}}\right)$. This allows us to choose $x \in C^{\geq \frac{\pi}{2}}$ and $y \in \left(C^{\geq \frac{\pi}{2}}\right)^{\geq \frac{\pi}{2}}$ such that $|px|, |py| < \frac{\pi}{2}$. If $\angle\left(\overrightarrow{px}, \overrightarrow{py}\right) \leq \frac{\pi}{2}$, then angle comparison implies that $|xy| < \frac{\pi}{2}$ a contradiction. □

Lemma 5.3 *Let X be an Alexandrov space with* curv ≥ 1. *If $\partial X \neq \emptyset$, then X has a unique soul $s \in X$ at maximal distance from the boundary,* rad $X \leq \frac{\pi}{2}$, *X is homeomorphic to the cone over the boundary, and $S_s X$ is homeomorphic to ∂X.*

Proof The first part of the proof is identical to the first part of the proof of Theorem 4.8. The function $d_1(x) = \sin|x\partial X|$ satisfies $d_1'' \leq -d_1$. Consider a unit speed geodesic $c(t) : [0, L] \to X$ from $x \in \partial X$ to a point $s \in X$ that is at maximal distance from the boundary. If $|sx| > \frac{\pi}{2}$, then $(d_1 \circ c)'(L) = \cos L < 0$ which shows that $d_1 \circ c$ had to have a maximum prior to reaching s, a contradiction. This shows that $X \subset \bar{B}\left(s, \frac{\pi}{2}\right)$. In particular, d_1 is strictly concave on $X - \partial X$, which implies that s is the unique point at maximal distance.

The topological statements follow from the fibration theorem. We start by noting that $r(x) = |x\partial X|$ defines a proper map from $X - \{s\}$ to $[0, |s\partial X|)$ that has no critical points. To see why the latter holds select a unit speed geodesic $c(t) : [0, b] \to X$ from $x \in X - \partial X$ to s. If x is critical for r, then some minimal geodesic from x to ∂X will have angles $\leq \frac{\pi}{2}$ with c. This shows that $(d_1 \circ c)'(0) \leq 0$. As also $d_1'' \leq 0$, it follows that d_1 is nonincreasing and consequently that $x = s$. The fibration theorem now implies that $X - \{s\} \simeq [0, |s\partial X|) \times \partial X$ and thus that X is a cone over the boundary.

The second topological claim uses a similar but slightly more complicated construction. From [17] one obtains a strictly concave function g near s which has unique maximum at s and whose rescaled level sets are homeomorphic to $S_s X$. On the other hand by Perel'man's fibration theorem we have that the level sets of r that are near the soul are homeomorphic to ∂X. The interpolated functions $r_\epsilon = (1 - \epsilon) r + \epsilon g$ are also strictly concave near s with a unique maximum at s for all $\epsilon \in [0, 1]$. The function $(x, \epsilon) \mapsto (r_\epsilon(x), \epsilon)$ is both tight and has no critical points on a suitable deleted neighborhood around s. Moreover, the level sets sufficiently near s are compact and by the fibration theorem homeomorphic. Considering $\epsilon = 0, 1$ we see that the level sets of r near s are homeomorphic to $S_s X$. □

This lemma shows that $\text{gexp}_s(1; \cdot) : C_1(S_s X) \to X$ is onto. Since this map is also distance nonincreasing it follows that rad $S_s X \geq \frac{\pi}{2}$ when rad $X = \frac{\pi}{2}$ (see also Proposition 5.1).

In the extreme case where rad $X = \frac{\pi}{2}$ we start with the preliminary definition of the edge as the dual set to s:

$$P = \left\{x \in X \mid |xs| = \frac{\pi}{2}\right\}.$$

It has the following important property.

Proposition 5.4 *Let X be an Alexandrov space with* curv ≥ 1, $\partial X \neq \emptyset$, *and* rad $= \frac{\pi}{2}$. *If $\partial X \neq \emptyset$ and* rad $X = \frac{\pi}{2}$, *then s is at maximal distance from P.*

Proof We first show that when s is a critical point for the distance to P, then s is in fact at maximal distance from P. To see this select $x \in X$ and a geodesic direction $\overrightarrow{sx} \in S_s X$. If s is critical for P, then there is a geodesic direction $\overrightarrow{se} \in S_s X$ with $e \in P$ such that $\angle\left(\overrightarrow{sx}, \overrightarrow{se}\right) \leq \frac{\pi}{2}$. Toponogov comparison then implies that $|xe| \leq \frac{\pi}{2}$ as $|xs|, |es| \leq \frac{\pi}{2}$.

Thus we need to show that when s is not critical, then rad $X < \frac{\pi}{2}$. For that choose a unit speed geodesic $c(t)$ such that $\dot{c}(0) \in S_s X$ forms an angle $< \frac{\pi}{2} - \epsilon$ with every direction $\overrightarrow{se} \in S_s X$ and $e \in P$. We can now find an open neighborhood $U \supset P$ such that the directions $U' \subset S_s X$ for minimal geodesics from s to points in U also form an angle $< \frac{\pi}{2} - \epsilon$ with $\dot{c}(0)$. By compactness there exists $\delta > 0$ such that $|sz| \leq \frac{\pi}{2} - \delta$ for all $z \notin U$. So we can in addition fix t such that $|c(t)z| \leq \frac{\pi}{2} - \frac{\delta}{2}$ for all $z \notin U$. On the other hand if $\overrightarrow{sy} \in S_x X$ denotes a direction to a $y \in U$, then Toponogov comparison implies

$$\cos|c(t)y| \geq \cos t \cos|sy| + \sin t \sin|sy| \cos\left(\frac{\pi}{2} - \epsilon\right).$$

Here the left-hand side is uniformly positive for any fixed small t, so there is an $\epsilon_1 > 0$, such that $|c(t)y| \leq \frac{\pi}{2} - \epsilon_1$ for all $y \in U$. This shows that rad $X < \frac{\pi}{2}$. □

Proposition 5.5 *Let X be an Alexandrov space with* curv ≥ 1, $\partial X \neq \emptyset$, *and* rad $= \frac{\pi}{2}$. *It follows that*

(1) *the gradient curves for $r(x) = |x\partial X|$ that start in P are minimal geodesics from P to s,*
(2) rad $P \geq \frac{\pi}{2}$,
(3) *when* dim $P = 0$, *then* $X = \Sigma_1 S$.

Proof Let $c : [0, L] \to X$ be a gradient curve for r reparametrized by arclength with $c(0) \in \partial X$ and $c(L) = s$. We will use comparison along this gradient curve as in [27, Lemma 2.1.3]. To that end consider $d_1 = \sin(r \circ c)$ and note that c is clearly also a gradient curve for d_1. This implies $d_1'' + d_1 \leq 0$ in the support sense. Since $d_1(0) = 0$ this shows that $d_1(t) \leq d_1'(0) \sin t$. As $d_1 \geq 0$, this implies that $L \leq \pi$. Define $g(\tau) = d_1(L - \tau)$ and note that also $g'' + g \leq 0$ in the support sense. This time

$$g(\tau) \leq g(0) \cos\tau + g'(0) \sin\tau.$$

At $\tau = L$ this becomes

$$0 \leq g(0) \cos(L) + g'(0) \sin(L),$$

where $g(0) > 0$ and $g'(0) \leq 0$ as r, d_1, and hence g are maximal at s. Since $L \leq \pi$, this forces $L \leq \frac{\pi}{2}$. On the other hand we always have $L \geq |sc(0)|$ so when $c(0) = e \in P$ this shows that $L = \frac{\pi}{2}$ and that the gradient curve must be a minimal geodesic from e to s. This proves (1).

For (2) we use that P consists of all points at distance $\frac{\pi}{2}$ from s, and hence is a convex subset of X. Moreover, by proposition 4.2 intrinsic distances in P are the same as the extrinsic distances in X. Assume that $B(P, \epsilon) \subset B\left(e, \frac{\pi}{2}\right)$. In (1) we saw that there is a minimal geodesic $c : \left[0, \frac{\pi}{2}\right] \to X$ from s to e which is a reparametrized gradient curve for r. We first claim that $\overrightarrow{es} = \dot{c}^{-}\left(\frac{\pi}{2}\right) \in S_eX$ is the soul. In fact, by first variation, $d_e r(\xi)$, $\xi \in S_eX$, is maximal when ξ is the soul of S_eX (see [27, Definition 1.3.2]). Thus, $\angle\left(w, \overrightarrow{es}\right) \leq \frac{\pi}{2}$ for all $w \in S_eX$ and by Toponogov comparison $B(P, \epsilon) \subset B\left(c(t), \frac{\pi}{2}\right)$ for all $t > 0$. However, also $X - B(P, \epsilon) \subset B\left(c(t), \frac{\pi}{2}\right)$ for sufficiently small t. This shows that $\operatorname{rad} X < \frac{\pi}{2}$.

Finally (3) follows from the above. Since P is π-convex and has $\operatorname{rad} \geq \frac{\pi}{2}$ it follows that $P = \{0, \pi\}$. Thus $\operatorname{diam} X = \operatorname{diam} P = \pi$ and the result follows. □

Define $P' \subset S_sX$ as the directions $\overrightarrow{se}$, $e \in P$, that correspond to the gradient curves for r as in part (1) of Proposition 5.5. Note that we have not excluded the possibility that there might be other minimal geodesics from s to points in P that do not correspond to such gradient curves.

Corollary 5.6 *The spherical gradient exponential gives a distance nonincreasing homeomorphism from P' to P.*

Proof If we think of the spherical exponential map as a map $\operatorname{gexp}(1, \cdot) : C_1(S_sX) \to X$, then $\partial X \subset \operatorname{gexp}(1, S_sX)$ and $P = \operatorname{gexp}\left(1, P'\right)$. Since P' is compact this yields a distance nonincreasing homeomorphism $\operatorname{gexp}(1, \cdot) : P' \to P$. □

In case P has boundary (see example 3.15) it is convenient to redefine it to an edge $E \subset P$ which otherwise has similar properties. When $\partial P \neq \emptyset$, then it follows from part (2) of Proposition 5.5 that P itself has an edge $P_1 \subset P$. Continuing in this fashion we will reach a π-convex subset $E \subset P$ that has $\operatorname{rad} \geq \frac{\pi}{2}$ and empty boundary. Proposition 5.5 and Corollary 5.6 clearly also hold for E in place of P.

We define the *spine* as the dual set to E by

$$S = \left\{x \in X \mid |xE| \geq \frac{\pi}{2}\right\}.$$

Note that $s \in S$ and so E is the dual set to S. The following corollary is now a consequence of proposition 4.2.

Proposition 5.7 *Assume that X has* curv ≥ 1, $\partial X \neq \emptyset$, *and* $\operatorname{rad} X = \frac{\pi}{2}$. *It follows that*

(1) *S is closed and π-convex in X.*
(2) *E, respectively, S is a deformation retract of $X - S$, respectively, of $X - E$*
(3) *E, respectively, $S \cap \partial X$ is a deformation retract of $\partial X - S \cap \partial X$, respectively, of $\partial X - E$*

Concrete deformations are provided by the gradient flows for the distance functions to S and E, respectively, preserving the extremal set ∂X.

We will now see that $E \neq \partial X$ if and only if $\dim S > 0$, in which case $\partial S \neq \emptyset$.

Lemma 5.8 *Assume that X has $\operatorname{curv} \geq 1$, $\partial X \neq \emptyset$, and $\operatorname{rad} X = \frac{\pi}{2}$.*

(1) *If $E \neq \partial X$, $x \in S$, and $q \in \partial X$ is closest to x, then $q \in S$. In particular, $\partial S \neq \emptyset$.*
(2) *The distance function to $S \cap \partial X$ on S is strictly concave and has its maximum at s. In particular, S is homeomorphic to the cone on $S \cap \partial X$. Moreover, s is the soul of S if and only if $S \cap \partial X = \partial S$, and if not $s \in \partial S$.*
(3) *When $\dim S = 0$, then $E = \partial X$ and $X = C_1 E = C_1 \partial X$.*

Proof

(1) The choice of q combined with Remark 4.2 shows that $\angle\left(\overrightarrow{qx}, \overrightarrow{qe}\right) \leq \frac{\pi}{2}$ for any $\overrightarrow{qe}$, $e \in E$. Since $|xe| \geq \frac{\pi}{2}$ angle comparison implies
$$0 \geq \cos|xe| \geq \cos|xq|\cos|eq|.$$
As $|xq| < \frac{\pi}{2}$ this shows that $|eq| \geq \frac{\pi}{2}$ and $q \in S$. In particular, the closest points in ∂X to the soul s lie in S and hence in ∂S.
(2) We know that S is π-convex and by (1): $|x\,\partial X| = |x\,S \cap \partial X|$ for all $x \in S$. This proves all but the last claim. When $S \cap \partial X \neq \partial S$, then $\partial S = (S \cap \partial X) \cup (\partial S \cap \operatorname{int} X)$. In this case, $S \cap \partial X$ is a face of the boundary ∂S, and s is the soul point of S relative to this face.
(3) Proposition 5.7 shows that any gradient curve for E reaches S. If we start in a direction of ∂X, then the gradient curve will stay in the extremal set ∂X. Therefore, it reaches a point in $\partial X \cap S$. This is clearly not possible when $S = \{s\}$, so it follows that $E = \partial X$.

Next we give the details of the rigidity statement. Consider $d_1(x) = \sin|x\partial X|$ and $f_1(x) = 1 - \cos|xs|$. These functions satisfy $d_1''(x) \leq -\sin|x\partial X|$ and $f_1''(x) \leq \cos|xs|$. Since every point on the boundary is at distance $\frac{\pi}{2}$ from s, it follows that $|x\partial X| + |xs| \geq \frac{\pi}{2}$. Hence,
$$\begin{aligned} d_1''(x) + f_1''(x) &\leq -\sin|x\partial X| + \cos|xs| \\ &\leq -\sin|x\partial X| + \cos\left(\frac{\pi}{2} - |x\partial X|\right) \\ &= 0. \end{aligned}$$

On the other hand

$$\begin{aligned} d_1(x) + f_1(x) &= \sin|x\partial X| + 1 - \cos|xs| \\ &\geq \sin|x\partial X| + 1 - \cos\left(\frac{\pi}{2} - |x\partial X|\right) \\ &= 1 \end{aligned}$$

with equality holding for any x that lies on a geodesic from s to ∂X. The minimum principle then shows that $d_1 + f_1 = 1$ on all of X. This shows that the gradient exponential map $\mathrm{gexp}_s(1;\cdot) : C_1(S_s X) \to X$ is an isometry.

□

Remark 5.9 This shows that $S \cap \partial X$ is nonempty provided $E \neq \partial X$ but not that $\partial S \subset \partial X$. In example 3.17 the soul $s \in \partial S$. In particular, s need not be the soul of S.

5.2 Proofs of Rigidity Results

The rigidity theorems from Sect. 4 depend on the following version of Lefschetz duality.

Theorem 5.10 *Assume Z is a compact connected ANR that is a $\mathbb{Z}_2$-homology sphere, and $A, B \subset Z$ are disjoint, compact, connected, and ANR. If $B \subset Z - A$ is a deformation retract and $Z - A$ is a topological n-manifold, then*

$$H_q(B; \mathbb{Z}_2) \simeq H^{n-1-q}(A; \mathbb{Z}_2), \; q = 1, \ldots, n-2.$$

Proof It follows from Lefschetz duality and the fact that A and Z are ANRs that for all q

$$H_q(Z - A; \mathbb{Z}_2) \simeq H^{n-q}(Z, A; \mathbb{Z}_2).$$

The assumption that $H^p(Z; \mathbb{Z}_2) = 0$ for $p = 1, \ldots, n-1$ shows, via the long exact sequence for relative cohomology, that

$$H^{n-q}(Z, A; \mathbb{Z}_2) \simeq H^{n-1-q}(A; \mathbb{Z}_2), q = 1, \ldots, n-2.$$

Finally, as $B \subset Z - A$ is a deformation retract the claim follows. □

We require some extra notation. For a convex subset $A \subset X$ of an Alexandrov space we define the normal space at $a \in A$ as the dual space to $S_a A \subset S_a X$:

$$N_a A = \left\{ v \in S_a X \mid \angle (v, w) \geq \frac{\pi}{2} \text{ for all } w \in S_a A \right\}.$$

By first variation any unit speed geodesic that starts in a and minimizes the distance to A has initial velocity that lies in $N_a A$.

Lemma 5.11 *Assume $Z = \bar{Z}/H$ is an Alexandrov space with* curv ≥ 1 *and* $\partial Z = \emptyset$*, where H is a finite group of isometries, $\bar{Z}$ is a closed topological n-manifold that is a $\mathbb{Z}_2$-homology sphere, and an Alexandrov space with* curv ≥ 1*. If $A, B \subset Z$ form a dual pair and $\partial A = \emptyset$, then there exist $x \in B$, $y \in A$, and finite group, G, that acts effectively and isometrically on both $N_x B$ and $N_y A$, such that $Z = \left(N_x B * N_y A\right)/G$.*

Proof The goal is to prove that

$$\dim A + \dim B = n - 1$$

and that no points in B have distance $> \frac{\pi}{2}$ to A and vice versa. This allows us to use [29, Theorem A] when $\partial B = \emptyset$ and otherwise [29, Theorem B] to reach the conclusion of the lemma.

We lift the situation to $\bar{A}, \bar{B} \subset \bar{Z}$. In case $\bar{A}$ (or $\bar{B}$) is not connected the components must be distance π apart as $\bar{A}$ is π-convex. This forces $\bar{A}$ to consist of two points and $\bar{Z}$ to be a suspension with $\bar{B} = S_a X$, $a \in \bar{A}$. So we can assume that $\dim \bar{A} = p > 0$. Since $\bar{A}, \bar{B}$ form a dual pair it follows that each of these sets contains the set of critical points for the distance function to the other set. The gradient flow then shows that $\bar{Z} - \bar{A}$ deformation retracts to $\bar{B}$. From $\partial \bar{A} = \emptyset$ we conclude that $H^p\left(\bar{A}, \mathbb{Z}_2\right) = \mathbb{Z}_2$. By Alexander duality we can then conclude that $H_q\left(\bar{B}, \mathbb{Z}_2\right) = \mathbb{Z}_2$ for $p = n - 1 - q$. This shows that

$$\dim \bar{A} + \dim \bar{B} \geq n - 1.$$

Since $\bar{B} \subset \bar{Z}$ is convex it follows that $\partial \bar{B} = \emptyset$ as it would otherwise be contractible. Frankel's theorem for Alexandrov spaces (see [26]) then shows that

$$\dim \bar{A} + \dim \bar{B} \leq n - 1.$$

Moreover, points in $\bar{B}$ cannot have distance $> \frac{\pi}{2}$ from $\bar{A}$ and vice versa. This finishes the proof. □

Remark 5.12 It is in general not possible to conclude that $\bar{Z}$ in Lemma 5.11 is a join. The icosahedral group I acts freely on $S^3(1)$ and hence on $S^3(1) * S^3(1) = S^7(1)$. While the quotient $S^7(1)/I$ is clearly a homology sphere it is not a join as it is a space form that is not homeomorphic to a sphere.

Remark 5.13 **Note that from the classification obtained in** [9] any positively curved Alexandrov space of dimension ≤ 3 and empty boundary is of the form $Z = \bar{Z}/H$ where $\bar{Z}$ is homeomorphic to a sphere. If in addition $\operatorname{diam} Z = \frac{\pi}{2}$, then we obtain two dual sets $A, B \subset Z$. If both of these have boundary, then Z is topologically a suspension and therefore topologically a sphere or $\Sigma\mathbb{RP}^2$. Otherwise we can apply the lemma.

We can now prove a topological regularity theorem that leads to the proofs of the other rigidity results.

Theorem 5.14 *Let X^n be an Alexandrov space with* curv ≥ 1, $\partial X \neq \emptyset$, *and* $\operatorname{rad} X = \frac{\pi}{2}$. *If ∂X is a topological manifold and a $\mathbb{Z}_2$-homology sphere, then $D(X) = (X_1 * X_2)/G$. Here G is a finite group acting effectively and isometrically on both X_1 and X_2 whose action is extended to the spherical join $X_1 * X_2$.*

Proof The idea is simply to apply Lemma 5.11 to the double. This requires a few minor adjustments. We use the dual decomposition for $D(X)$ that consists of $D(S)$ and the copy of $F \subset D(X)$ that corresponds to $E \subset \partial X$. Here $D(S)$ will turn out to be the double of S, but for now it is simply the preimage of S. Note that inside X **the gradient flows for** S and E preserve ∂X. Thus we obtain deformation retractions of $D(X) - D(S)$ to F and $D(X) - F$ to $D(S)$ relative to ∂X as in Proposition 5.7. Moreover, as X is homeomorphic to the cone over the boundary it follows that $D(X) - D(S)$ is a topological n-manifold. Additionally, $D(X)$ is a $\mathbb{Z}_2$-homology sphere by Meyer-Vietoris. This again shows that

$$H_q(F; \mathbb{Z}_2) \simeq H^{n-1-q}(D(S); \mathbb{Z}_2), \ q = 1, \ldots, n-2.$$

We can then argue as in the proof of Lemma 5.11 that $\partial D(S) = \emptyset$ and that we obtain the desired decomposition for $D(X)$. □

Remark 5.15 Note that $N_x F \subset S_x D(X)$ is the double of $N_x E \subset S_x X$. This shows that $N_x F$, and thus also $N_y D(S) * N_x F$, **come with a natural reflection whose** quotient is $N_x E$, respectively, $N_y D(S) * N_x E$. Let R **be the natural reflection on** both $N_x F$ and $D(S) = N_x F/G$. This results in a commutative diagram

$$\begin{array}{ccc} N_x F & \xrightarrow{R} & N_x F \\ \downarrow & & \downarrow \\ D(S) & \xrightarrow{R} & D(S). \end{array}$$

In case G commutes with R on $N_x F$ it follows that G will also act on $N_x E$. Consequently, also X becomes the quotient of a join: $X = (N_x E * N_y S)/G$. It is not, in general, clear whether R commutes with elements in G. However, in the case where $G = \langle g \rangle$, $g^2 = \mathrm{id}$ this is automatically true. To see this assume $z, g(z) \in N_x F$ are mapped to $s \in D(S)$. Then $R(z), R(g(z))$ are both mapped to

$R(s)$, but the preimage of $R(s)$ also consists of the two points $R(z), g(R(z))$, this shows that $gR = Rg$.

We can now prove the main result.

Theorem 5.16 *Let X^n be an Alexandrov space with* curv ≥ 1, $\partial X \neq \emptyset$, *and* $\operatorname{rad} X = \frac{\pi}{2}$. *If* $\operatorname{rad} S_s X > \frac{\pi}{2}$, *then* $X = E * S$, *where* $\operatorname{rad} E > \frac{\pi}{2}$ *and* $\operatorname{rad} S_s S > \frac{\pi}{2}$.

Proof The radius assumption on the space of directions is first used to see that X is a topological manifold near the soul. Since X is homeomorphic to a cone it follows that it is a topological manifold with boundary. Finally, the space of directions is homeomorphic to a sphere (see [11] and [27]) and thus ∂X is also homeomorphic to a sphere. This shows that we can apply Theorem 5.14.

Next we can use it in combination with the following fact: If G is a compact group acting by isometries on an Alexandrov space Y with $\operatorname{curv} Y \geq 1$, then $\operatorname{rad}(Y/G) \leq \frac{\pi}{2}$ unless the action is trivial. Moreover, if the action is free, then $\operatorname{diam} Y/G \leq \frac{\pi}{2}$. This is obvious when $\dim Y = 1$ and follows in general from an induction argument. To see this, first note that when $\operatorname{diam}(Y/G) > \frac{\pi}{2}$, then there are two orbits that are at distance $> \frac{\pi}{2}$ from each other. Thus one orbit is forced to lie in a π-convex set that is at distance $> \frac{\pi}{2}$ from some point. The action must then have a fixed point y in this set. It now follows from induction that when the action is nontrivial, then $\operatorname{rad}\left(S_y Y/G\right) \leq \frac{\pi}{2}$, and hence that $\operatorname{rad}(Y/G) \leq \frac{\pi}{2}$ (see e.g. [27, Lemma 5.2.1] and Proposition 5.1).

With notation as in the proof of Theorem 5.14 we show that $D(X) = F * D(S)$. Let $s \in D(X)$ be fixed to be one of the two soul points and note that $S_s S = S_s D(S)$ and $S_s X = S_s D(X)$.

First we use Corollary 5.6 to show that $E' = N_s S$. Observe that as $E' \subset N_s S$ we have:

$$\dim N_s S + \dim S_s S \geq \dim E + \dim S - 1 = n - 2.$$

On the other hand as $\partial S_s S = \emptyset$ it follows from [29, theorem A, part (A1)] that

$$\dim N_s S + \dim S_s S \leq n - 2.$$

Thus $\dim E' = \dim N_s S$. In case both spaces are 0-dimensional they will both consist of two points at distance π apart. This is because $\operatorname{rad} E' \geq \frac{\pi}{2}$ while $N_s S \subset S_s X$ is π-convex. When both spaces have dimension > 0, we know that $N_s S$ is connected as it is π-convex. Since $\dim E' = \dim N_s S$ and $\partial E' = \emptyset$ it follows that $E' \subset N_s S$ is an open and closed subset and hence that $E' = N_s S$.

This shows that only one point $\bar{s} \in N_x F \subset N_y D(S) * N_x F$ is mapped to $s \in D(S) \subset D(X)$ and hence that $\bar{s}$ is a fixed point for the isometric action of G on $N_y D(S) * N_x F$. In particular, G preserves $S_{\bar{s}}\left(N_y D(S) * N_x F\right)$ and $S_s X = \left(S_{\bar{s}}\left(N_y D(S) * N_x F\right)\right)/G$. However, this can only happen if G acts trivially on $S_{\bar{s}}\left(N_y D(S) * N_x F\right)$ as $\operatorname{rad} S_s X > \frac{\pi}{2}$. In conclusion, G acts trivially on $N_y D(S) *$

$N_x F$ and $D(X)$ is a spherical join. This shows that $S_s X = E' * S_s S$ and rad $E' > \frac{\pi}{2}$ and rad $S_s S > \frac{\pi}{2}$.

Finally note that the natural reflection on $D(X) = D(S) * F$ fixes F and has orbit space X. Thus also $X = S * E$. □

Remark 5.17 In the context of this result it is worth noting that the icosahedral group I acts on $S^5(1) = S^1(1) * S^3(1)$ with a quotient $S^1(1) * \left(S^3(1)/I\right)$ that is a topological sphere (see e.g., [5]). This quotient also shows that one can not expect to use general position or transversality arguments for Alexandrov spaces that are topological manifolds.

Based on Lemma 5.11 we obtain the following generalization of Theorem 5.14.

Corollary 5.18 *Let X^n be an Alexandrov space with* curv ≥ 1, $\partial X \neq \emptyset$, *and* rad $X = \frac{\pi}{2}$. *If $S_s X$ is a topological manifold that is covered by a $\mathbb{Z}_2$-homology sphere or $S_s X$ is homeomorphic to $\Sigma\mathbb{RP}^2 = S^3/\mathbb{Z}_2$, then $D(X) = (X_1 * X_2)/G$. Here G is a finite group acting effectively and isometrically on both X_1 and X_2 whose action is extended to the spherical join $X_1 * X_2$.*

Proof The goal is to find a suitable ramified or branched cover of X. In all cases these will in fact be good orbifold covers. Specifically, for a given Alexandrov space X we seek an Alexandrov space Y, with the same lower curvature bound, and a finite group G acting by isometries on Y such that $X = Y/G$. The relevant results are established in [15, Theorem A] and [6, Sect. 2.2]. The main technical tools for showing that Y has the desired properties are proven in [20]. We only need these covers for positively curved spaces with boundary, i.e., for spaces that are homeomorphic to cones over the space of directions at the soul $X \simeq C_1 S_s X$. This means that the cover $Y \simeq C_1 S_{\bar{s}} Y$ and $(S_{\bar{s}} Y)/G = S_s X$ since G is forced to fix $\bar{s}$. Here $S_{\bar{s}} Y$ is either a covering space over $S_s X$ or a good orbifold cover of $S_s X$.

When $S_s X$ has a covering space there is only one isolated branch point and the complement of the soul is convex. This means we can use exactly the same strategy as in [6, Sect. 2.2] to create the Alexandrov space structure on Y. When $S_s X$ is homeomorphic to $\Sigma\mathbb{RP}^2$ it follows that X is not orientable as it does not have a local orientation at the soul. This means that we can use the orientation covering as in [15, Theorem A] as Y.

The assumptions of the corollary now show that Y exists and is a topological manifold. We can then apply Lemma 5.11 to finish the proof. □

Corollary 5.19 *When* dim $X \leq 4$, *then X is either isometric to a join or a $\mathbb{Z}_2$ quotient of a join, where $\mathbb{Z}_2$ acts effectively on both factors of the join.*

Proof Note that when dim $X \leq 2$, then either dim $E = 0$ or dim $S = 0$. Thus the result follows from Proposition 5.5 and Lemma 5.8. Similarly, in higher dimensions we can assume that both S and E have positive dimensions.

In case dim $X = 3, 4$ we need to use Corollary 5.18 and Remark 5.15. First observe that when dim $X = 3$, it follows that $S_s X$ is homeomorphic to S^2 or $\mathbb{RP}^2$, while when dim $X = 4$ we can use the classification from [9] to conclude that $S_s X$

is homeomorphic to a spherical space form or the suspension over the real projective plane. This places us in a position where we can use Corollary 5.18.

We assume that $D(X) = X_1 * X_2/G$, with $X_1/G = D(S)$ and $X_2/G = E$. When G is trivial there is nothing to prove so we also assume $|G| \geq 2$.

Since $\operatorname{rad} E \geq \frac{\pi}{2}$ and $\partial E = \emptyset$ we note that when $\dim X_2 = 1$ we have $G = \mathbb{Z}_2$ and acting as the antipodal map.

Assume that $\dim X_2 = 2$. When $\operatorname{diam} E = \pi$ we obtain a dimension reduction.

In case $\operatorname{diam} E \in \left(\frac{\pi}{2}, \pi\right)$ it follows as in the proof of Theorem 5.16 that G has two fixed points $x, y \in X_2$ with $|xy| = \operatorname{diam} X_2$. We then have that $S_{x,y}E = \left(S_{x,y}X_2\right)/G = S^1\left(\frac{1}{|G|}\right)$. Let $z \in E$ be any point with $|zx| = |zy| < \frac{\pi}{2}$. The radius assumption shows that there is $\bar{z} \in E$ with $|z\bar{z}| = \frac{\pi}{2}$. Consider a hinge with vertex at x (or y). Since $\operatorname{diam} S_{x,y}E \leq \frac{\pi}{2}$ the angle of the hinge is $\leq \frac{\pi}{2}$. But this leads to a contradiction as either $|x\bar{z}| < \frac{\pi}{2}$ or $|y\bar{z}| < \frac{\pi}{2}$ which by Toponogov comparison implies $|z\bar{z}| < \frac{\pi}{2}$. This shows that this situation is impossible.

Finally we consider the case when $\operatorname{diam} E = \operatorname{rad} E = \frac{\pi}{2}$. This allows us to obtain dual sets $A, B \subset E$ where, say, $\dim A = 0$. By Lemma 5.11 $E = \left(N_x B * N_y A\right)/\Gamma$, where Γ acts effectively on $N_x B$ and $N_y A$. Since $\dim N_x B = 0$ this implies that either $N_x B$ is a point and Γ is trivial or $N_x B$ consists of two points distance π apart and $\Gamma = \mathbb{Z}_2$. In the former case $E = \{0\} * S^1\left(\frac{1}{2}\right)$ which is impossible as $\partial E = \emptyset$. In the latter case $E = \left(\Sigma_1 S^1(r)\right)/\langle I\rangle$, $r \in \left[\frac{1}{2}, 1\right]$, where I interchanges the two suspension points and is an involution on $S^1(r)$. Such an involution is either the antipodal map or a reflection. When I is a reflection it fixes two points at distance πr apart which implies that $\operatorname{diam} E = \pi r$ and consequently $r = \frac{1}{2}$. In E consider two points e, f that form an angle θ at A. By Toponogov comparison

$$\cos|ef| \geq \cos|Ae|\cos|Af| + \sin|Ae|\sin|Af|\cos\theta.$$

So when $\theta \leq \frac{\pi}{2}$ and $|Ae|, |Af| \in \left(0, \frac{\pi}{2}\right)$ we have $|ef| < \frac{\pi}{2}$. This shows that $\operatorname{rad} E < \frac{\pi}{2}$, when I is a reflection and $r = \frac{1}{2}$. When I is an antipodal map we must have $\theta = \pi r$ to obtain maximal distance between e and f. So when $|Ae|, |Af| \in \left(0, \frac{\pi}{2}\right)$ and $r < 1$ we have $|ef| < |Ae| + |Af|$ as well as $|ef| \leq \pi - |Ae| - |Af|$ when we measure the distance through B. But this also forces $|ef| < \frac{\pi}{2}$. Thus E is isometric to $\mathbb{RP}^2(1)$. This also forces $X_2 = S^2(1)$ with G acting as the antipodal map.

We can now use Remark 5.15 to conclude that either $X = (X_1/\langle R\rangle) * X_2$ or $X = ((X_1/\langle R\rangle) * X_2)/\mathbb{Z}_2$. □

Remark 5.20 If we combine the constructions in the proof and Remark 5.13, then it is possible to obtain a classification for Alexandrov spaces with curv ≥ 1, rad $\geq \frac{\pi}{2}$ in low dimensions. It would be interesting to investigate what results one can obtain for such spaces in higher dimensions when they are not finite quotients of spheres. Do they, e.g., admit submetries onto $\left[0, \frac{\pi}{2}\right]$?

References

1. S. Alexander and R. Bishop, *Extrinsic curvature of semiconvex subspaces in Alexandrov geometry*. Ann. Global Anal. Geom. 37 (2010), no. 3, 241–262.
2. S. Brendle, F. Marques, and A. Neves, *Deformations of the hemisphere that increase scalar curvature*. Invent. Math. 185 (2011), no. 1, 175–197.
3. D. Burago, Yu. Burago, and S. Ivanov, *A Course in Metric Geometry*, Graduate Studies in Math. Vol 33. AMS, 2001.
4. Yu. Burago, M. Gromov, and G. Perel'man, *Aleksandrov Spaces with Curvatures Bounded from Below*. Russian Math. Surveys 47 (1992), no. 2, 1–58.
5. R. J. Daverman, *Decompositions of Manifolds*, AMS Chelsea Publishing 2007.
6. Q. Deng, F. Galaz-Garcia, L. Guijarro, M. Munn, *Three-Dimensional Alexandrov Spaces with Positive or Nonnegative Ricci Curvature*, Potential Anal. (2018) 48:223–238.
7. J. Ge and R Li, *Radius estimates for Alexandrov space with boundary*, https://arxiv.org/abs/1812.02571
8. J. Ge and R. Li, *Rigidity for positively curved Alexandrov spaces with boundary*, https://arxiv.org/abs/1811.04257
9. F. Galaz-Garcia and L. Guijarro, *On Three-Dimensional Alexandrov Spaces*. International Mathematics Research Notices, Vol. 2015, No. 14, pp. 5560–5576.
10. K. Grove and D. Gromoll, *A generalization of Berger's rigidity theorem for positively curved manifolds*, Ann. Sci. École Norm. Sup. (4), 20 (1987), no.2, 227–239.
11. K. Grove and P. Petersen, *A radius sphere theorem*, Invt. Math. 112, 577–583 (1993).
12. K. Grove and P. Petersen, *Alexandrov Spaces with maximal radius*, ArXiv 1805.10221 v3, to appear in Geometry & Topology.
13. K. Grove, A Moreno, and P. Petersen, *The Boundary Conjecture for Leaf Spaces*, preprint arXiv 1804.01656.
14. F. Hang and X. Wang, *Rigidity theorems for compact manifolds with boundary and positive Ricci curvature*. J. Geom. Anal. 19 (2009), no. 3, 628–642.
15. J. Harvey and C. Searle, *Orientation and Symmetries of Alexandrov Spaces with Applications in Positive Curvature*, J Geom Anal (2017) 27:1636–1666.
16. W. Hurewicz and H. Wallman, *Dimension Theory*, Princeton University Press 1941.
17. V. Kapovitch, *Regularity of limits of noncollapsing sequences of manifolds*. Geom. Funct. Anal. 12 (2002), no. 1, 121–137.
18. V. Kapovitch, *Perel'man's stability theorem*. Surveys in differential geometry. Vol. XI, 103–136, Surv. Differ. Geom., 11, Int. Press, Somerville, MA, 2007.
19. K. Kuratowski, *Set Theory*, North-Holland, Amsterdam, (1976).
20. N. Li, *Globalization with probabilistic convexity*, J. Top. Anal. vol. 7 no. 4 (2015) 719–735.
21. A. Lytchak, *Open map theorem for metric spaces*. Algebra i Analiz 17 (2005), no. 3, 139–159; reprinted in St. Petersburg Math. J. 17 (2006), no. 3, 477–491
22. P. Miao, *Positive mass theorem on manifolds admitting corners along a hypersurface*, Adv. Theo. Math. Phys. 6 (2002).
23. M. Min-Oo, *Scalar curvature rigidity of asymptotically hyperbolic spin manifolds*. Math. Ann. 285, 527–539 (1989).
24. G. Perel'man, *Alexandrov's Spaces with Curvatures Bounded from Below*, II.
25. P. Petersen, *Riemannian Geometry*, 3rd Ed. Graduate Texts in Mathematics 171. Springer, 2016.
26. A. Petrunin, *Parallel Translation for Alexandrov Space with Curvature Bounded Below*, GAFA vol. 8 (1998) 123–148.
27. A. Petrunin, *Semiconcave Functions in Alexandrov's Geometry*, arXiv:1304.0292 and Surveys in Differential Geometry. Vol. XI, 137–201, Int. Press, Somerville, MA, 2007.
28. C. Plaut, *Metric Spaces of Curvature* $\geq k$, Handbook of Geometric Topology, Chapter 16, Edited by R.J. Daverman and R.B. Sher, 2002 Elsevier Science B.V.

29. X. Rong and Y Wang, *Finite Quotient of Join in Alexandrov Geometry*, https://arxiv.org/abs/1609.07747
30. B. Wilking, *Index parity of closed geodesics and rigidity of Hopf fibrations*, Invent. Math. 144 (2001), no. 2, 281–295.

Three-Dimensional Alexandrov Spaces: A Survey

Fernando Galaz-García and Jesús Núñez-Zimbrón

2010 Mathematics Subject Classification 53C23, 52C20

1 Introduction

The natural objects of study in Riemannian geometry are smooth manifolds which carry a smooth Riemannian metric, that is, smooth Riemannian manifolds. Many useful tools have been developed to study these objects, including highly developed theories of geometric and functional analysis on Riemannian spaces. Riemannian manifolds also carry a natural metric space structure allowing not only for local analytic arguments, but also for global results linking geometry and topology.

On account of being metric spaces, compact Riemannian manifolds naturally fall in the context of the Gromov–Hausdorff distance, which makes the collection of (isometry classes of) compact metric spaces into a metric space. It is then natural to ask what the metric closure of the class of compact Riemannian manifolds is under Gromov–Hausdorff convergence. The answer to this question was obtained by Cassorla in [16]: this closure consists of all compact inner metric spaces.

Supported in part by research grants MTM2014-57769-3-P and MTM2017-85934-C3-2-P from the Ministerio de Economía y Competitividad de Españ (MINECO), by the DFG grant GA 2050 2-1 within the Priority Program SPP 2026 "Geometry at Infinity", and by the RTG 2229 "Asymptotic Invariants and Limits of Groups and Spaces" at Heidelberg University and the Karlsruhe Institute of Technology (KIT).

F. Galaz-García (✉)
Department of Mathematical Sciences, Durham University, Durham, UK
e-mail: fernando.galaz-garcia@durham.ac.uk

J. Núñez-Zimbrón
Centro de Investigación en Matemáticas (CIMAT), Guanajuato, Mexico
e-mail: jesus.nunez@cimat.mx

G. Arizmendi Echegaray et al. (eds.), *Recent Advances in Alexandrov Geometry*, CIMAT Lectures in Mathematical Sciences,
https://doi.org/10.1007/978-3-030-99298-9_2

On a different note, Riemannian manifolds can be studied via several fundamental invariants, the most important being the curvature tensor. The curvature tensor itself gives rise to different notions of curvature (and, in particular, curvature bounds) such as *sectional*, *Ricci*, and *scalar* curvatures. One could then wonder about the interplay between curvature and Gromov–Hausdorff convergence, or particularly, what the closure of different classes of Riemannian manifolds with given curvature bounds is. This leads in a natural way to the consideration of different curvature notions on non-smooth metric spaces.

Motivated by the preceding considerations, much work has been devoted to extending the definitions of sectional, Ricci, and scalar curvature lower bounds to non-smooth metric spaces. By now, there exists a well established theory of metric spaces with sectional curvature bounded either above or below, known, respectively, as CAT [3, 10] or *Alexandrov* spaces [12, 13]. In the case of Ricci curvature bounded below, starting with the seminal work of Lott, Villani [62], and Sturm [95, 96], the theory of metric spaces with a lower curvature Ricci curvature has evolved into a well developed field that has seen much attention in recent years (see, for example, [36, 37]). Finally, a metric generalization of a uniform lower scalar curvature bound is, at the moment, open and remains the focus of current research (see, for example, [39, 40, 94, 97]).

Alexandrov spaces in particular play an important role in questions involving the global geometry of Riemannian manifolds and arise, for example, as Gromov–Hausdorff limits of convergent sequences of n-dimensional compact Riemannian manifolds with a uniform lower sectional curvature bound. Familiar examples of Alexandrov spaces include Riemannian orbifolds with a uniform lower sectional curvature bound and orbit spaces of isometric compact Lie group actions on complete Riemannian manifolds with sectional curvature bounded below. In addition to their relevance in Riemannian geometry, Alexandrov spaces are objects of interest in their own right and, since complete Riemannian manifolds with a uniform lower sectional curvature bound are Alexandrov spaces, Alexandrov geometry may be seen as a metric generalization of Riemannian geometry. Indeed, many Riemannian theorems, such as the Bonnet–Myers theorem, have corresponding analogues for Alexandrov spaces and much effort has been made in generalizing Riemannian results to Alexandrov spaces. This seemingly simple approach involves significant challenges, as in Alexandrov geometry one must make do without the smooth tools available for Riemannian manifolds, relying instead on purely metric machinery. Thus, in order to recover Riemannian results, one must usually devise new arguments depending only on metric considerations.

As noted above, generic Alexandrov spaces are not manifolds and their topological intricacies increase considerably with their dimension. Indeed, locally, an n-dimensional Alexandrov space is homeomorphic to a cone over an $(n-1)$-dimensional Alexandrov space with curvature bounded below by 1. The latter spaces are far from being classified, even in the Riemannian case. Moreover, even if the space is manifold, the set of metric singularities may be dense. It is natural then to first consider Alexandrov spaces of low dimensions. The topology and geometry of one- and two-dimensional Alexandrov spaces is, essentially, well understood (see for example [12, Corollary 10.10.3]). Therefore, the next step consists in analyzing three-dimensional Alexandrov spaces and in this survey we present an up-to-date and panoramic view of the topology and geometry of such spaces.

Our manuscript is organized as follows. In Sect. 2 we recall preliminary notions of Alexandrov geometry. In Sect. 3 we present the classification of positively and non-negatively curved spaces, the geometrization theorem, and a discussion on simply-connected and aspherical three-dimension Alexandrov spaces. We then present Lie group actions and their topological and equivariant classifications for three-dimensional Alexandrov spaces in Sect. 6. Finally, in Sect. 7 we give a brief account of results on collapsing Alexandrov spaces in dimension three.

2 Alexandrov Spaces

In this section we recall some basic concepts and general results on Alexandrov geometry. Most of this material can be found in the basic references [12, 13]; see also [83, 91]. We refer the reader to [1, 82] for further results on Alexandrov geometry as well as to Petersen's notes in this same volume.

2.1 *Basic Definitions*

Let (X, d) be a metric space. A *curve* (or *path*) in X is, by definition, a continuous function $\gamma\colon I \to X$ defined on an interval $I \subset \mathbb{R}$. We define the *length* of a curve $\gamma\colon [a, b] \to X$ by

$$\mathrm{Length}(\gamma) = \sup_{\mathcal{P}} \sum_{i=1}^{n} d(\gamma(t_{i-1}), \gamma(t_i)),$$

where the supremum is taken over all partitions of $[a, b]$, i.e., over all finite collections of points $\mathcal{P} = \{t_0, \ldots, t_n\}$ with $a = t_0 < t_2 < \cdots < t_n = b$. The metric space (X, d) is a *length space* if it is path connected and the distance $d(p, q)$ between points $p, q \in X$ is given by the infimum of lengths of curves joining p and q, i.e., curves $\gamma\colon [a, b] \to X$ with $\gamma(a) = p$ and $\gamma(b) = q$. In this case, d is said to be an *intrinsic* (or *inner*) metric. Note that any connected Riemannian manifold equipped with its Riemannian distance is a length space.

Let $I \subset \mathbb{R}$ be an interval. A curve $\gamma\colon I \to X$ is a *geodesic* if, for each interior value t of I, the restriction of γ to a small interval centered at t is a shortest path. A curve γ between points p and q in a metric space (X, d) is a *minimal geodesic* if $\mathrm{Length}(\gamma) = d(p, q)$. We will assume all geodesics to be minimal unless otherwise stated. It is worth noting that for locally compact and complete length spaces, one can always join every pair of points by a minimal geodesic [12, Theorem 2.5.23]. We will denote a geodesic between p and q by $[pq]$. Note that this terminology differs from the one used in Riemannian geometry, where a geodesic is a curve that locally minimizes distances between any two of its points and may not necessarily realize the distance between its endpoints. This may be easily verified by considering great circles on a round sphere.

Let us now define the *Hausdorff dimension* of a metric space Y. Let $\mathcal{V} = \{V_i\}_{i \in I}$ be a countable covering of Y, let $h \geq 0$ be a real number, and fix $\delta > 0$. Define

$$H_\delta^h(Y) = \inf_{\mathcal{V}} \left\{ \sum_{i \in I} \operatorname{diam}(V_i)^h : \operatorname{diam}(V_i) < \delta \text{ for all } i \in I \right\}, \tag{1}$$

where the infimum is taken over all countable coverings of Y by subsets of diameter less than δ. Note that if no such covering exists, then $H_\delta^h(Y) = \infty$. We convene that, if $h = 0$, then every 0^0 term appearing in the sum $\sum_{i \in I} \operatorname{diam}(V_i)^h$ in (1) is replaced by 1. We then define the *h-dimensional Hausdorff measure* of Y, denoted by $H^h(Y)$, by letting

$$H^h(Y) = C(h) \lim_{\delta \to 0} H_\delta^h(Y).$$

Here $C(h) > 0$ is a normalization constant chosen so that, if h is a positive integer, then $H^h([0, 1]^h) = 1$, where $[0, 1]^h$ is the unit cube in Euclidean space $\mathbb{R}^h$. We define the *Hausdorff dimension* of Y by

$$\dim_H(Y) = \inf \left\{ h \geq 0 : H^h(Y) = 0 \right\}.$$

We define the Hausdorff dimension of subsets of Y by considering such subsets as metric spaces equipped with the subspace metric induced by the metric on Y. Note that the Hausdorff dimension of a metric space may not necessarily be an integer. Indeed, many self-similar subspaces of Euclidean spaces, such as the Cantor set in $\mathbb{R}$ or the Sierpinski triangle in $\mathbb{R}^2$, have non-integer Hausdorff dimensions. On the other hand, the Hausdorff dimension of a Riemannian n-manifold equals its topological dimension.

We conclude this subsection by recalling the definitions of the *Hausdorff* and *Gromov–Hausdorff* distances. Let X be a metric space. Given $r > 0$ and a subset $S \subset X$, let $U_r(S) = \{ x \in X : d(x, S) < r \}$, i.e., $U_r(S)$ is an open r-neighborhood of S in X. The *Hausdorff distance* $d_H(A, B)$ between two subsets A, B of X is, by definition,

$$d_H(A, B) = \inf\{ r > 0 : A \subset U_r(B) \text{ and } B \subset U_r(A) \}.$$

Now, let X and Y be compact metric spaces. The *Gromov–Hausdorff* distance $d_{GH}(X, Y)$ between X and Y is given by

$$d_{GH}(X, Y) = \inf\{ d_H(f(X), g(Y)) \},$$

where the infimum is taken over all metric spaces Z and all isometric embeddings $f: X \to Z$ and $g: Y \to Z$. This distance measures how far X and Y are from being isometric. Indeed, $d_{GH}(X, Y) = 0$ if and only if X is isometric to Y. The space of (isometry classes of) compact metric spaces equipped with the Gromov–Hausdorff distance is itself a metric space.

From now on we will assume that (X, d) is a complete, locally compact length space. This guarantees the existence of geodesics between any two points in (X, d) (see [12, Theorem 2.5.23]). In the Riemannian setting the existence of shortest curves between any pair of points in a complete Riemannian manifold is a consequence of the Hopf–Rinow Theorem (see [22, Chap. 7, Theorem 2.8]). To lighten the notation, we will usually denote the metric space (X, d) simply by X. We will assume all our spaces to be connected.

2.2 Curvature Bounded Below

Rather than considering directly the geometry of a length space X, we will compare distances in X with distances in a given *model space* whose geometry is well understood. Our goal is to define a notion of (sectional) curvature bounded below using only distances in X.

The *model space* M_k^2 with curvature $k \in \mathbb{R}$ is the simply-connected complete Riemannian 2-manifold with constant sectional curvature k. Hence, M_k^2 is isometric to the Euclidean plane $\mathbb{R}^2$ if $k = 0$, to the round sphere of radius $1/\sqrt{k}$ if $k > 0$, or to the hyperbolic plane appropriately rescaled so that its sectional curvature is $k < 0$. Observe that

$$\operatorname{diam}(M_k^2) = \begin{cases} \pi/\sqrt{k} & \text{if } k > 0; \\ \infty & \text{if } k \leq 0. \end{cases}$$

A *geodesic triangle* Δpqr in X consists of three different points $p, q, r \in X$ and three geodesics $[pq]$, $[pr]$, $[qr]$ between them. Given $k \in \mathbb{R}$, a *comparison triangle for* Δpqr in M_k^2 is a geodesic triangle $\Delta\overline{pqr}$ in M_k^2 with vertices $\overline{p}, \overline{q}, \overline{r}$ and whose sides have the same length as the corresponding sides of Δpqr, i.e., $d(p, q) = d(\overline{p}, \overline{q})$, $d(p, r) = d(\overline{p}, \overline{r})$, and $d(q, r) = d(\overline{q}, \overline{r})$.

The angles of $\Delta\overline{pqr}$ at each one of its vertices $\overline{p}$, $\overline{q}$, and $\overline{r}$ are the *comparison angles* and we denote them, respectively, by $\tilde{\measuredangle}rpq$, $\tilde{\measuredangle}pqr$, and $\tilde{\measuredangle}prq$. Given a geodesic triangle Δpqr in X, a corresponding comparison triangle exists and is unique (up to an isometry of the model space) if $k \leq 0$, or if $k > 0$ and the perimeter of Δpqr is strictly less than $2\pi/\sqrt{k}$.

Suppose now that X is a complete Riemannian manifold with sectional curvature $\sec \geq k$ for some $k \in \mathbb{R}$. By Toponogov's comparison theorem, if Δpqr is a geodesic triangle in X and $\Delta\overline{pqr}$ is a comparison triangle in M_k^2, then

$$d(p, s) \geq d(\overline{p}, \overline{s}), \tag{2}$$

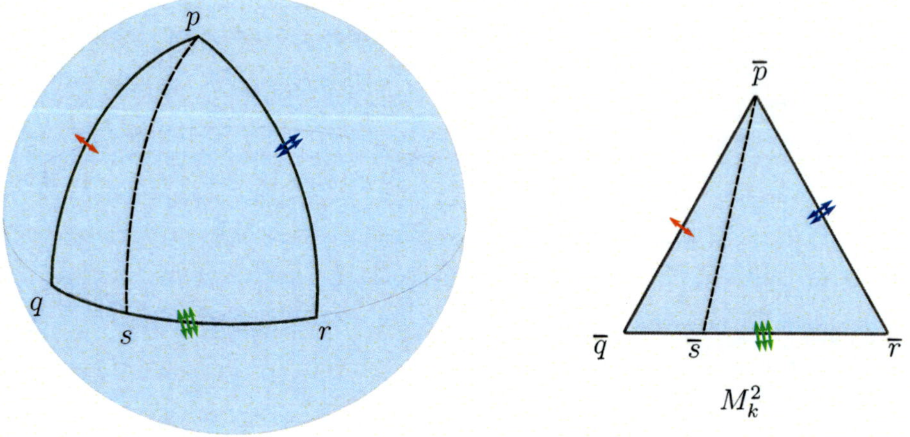

Fig. 1 The T_k-property. On the left hand side, Δpqr is displayed. On the right hand side, a comparison triangle $\Delta\overline{pqr}$ is displayed on M_k^2. **In the figure** M_k^2 has $k = 0$

where s is a point in the geodesic $[qr]$ in Δpqr and $\bar{s}$ is the point in the side $[\overline{qr}]$ of $\Delta\overline{pqr}$ with $d(\bar{s}, \bar{q}) = d(s, q)$. Conversely, if inequality (2) holds for any geodesic triangle Δpqr and a corresponding comparison triangle $\Delta\overline{pqr}$ in M_k^2, then X has sectional curvature bounded below by k (see [10, Theorem 1A.6]). We now take this metric characterization of sectional curvature bounded below as the definition of a lower (sectional) curvature for length spaces (Fig. 1).

Definition 2.1 (Property T_k**)** Let X be a complete, locally compact length space. A geodesic triangle Δpqr in X satisfies *property* T_k *for a given* $k \in \mathbb{R}$ if, for any comparison triangle $\Delta\overline{pqr}$ in M_k^2 and for any point s in the geodesic $[qr]$ in Δpqr,

$$d(p, s) \geq d(\overline{p}, \overline{s}),$$

where $\bar{s}$ is the point in the side $[\overline{qr}]$ of $\Delta\overline{pqr}$ with $d(\bar{s}, \bar{q}) = d(s, q)$.

Definition 2.2 A complete, locally compact length space X *has curvature bounded below by* $k \in \mathbb{R}$ (denoted by curv $\geq k$) if every point in X has an open neighborhood U_p where property T_k holds for every geodesic triangle in U_p.

With a metric definition of lower curvature bound now in hand, we define Alexandrov spaces.

Definition 2.3 A complete, locally compact length space of finite (Hausdorff) dimension is an *Alexandrov space with curvature bounded below by* $k \in \mathbb{R}$ if property T_k is satisfied locally.

Toponogov's globalization theorem [12, Theorem 10.3.1] implies that property T_k holds globally on any Alexandrov space with curvature bounded below by k, i.e., the property holds for any geodesic triangle in the space. We convene that the line

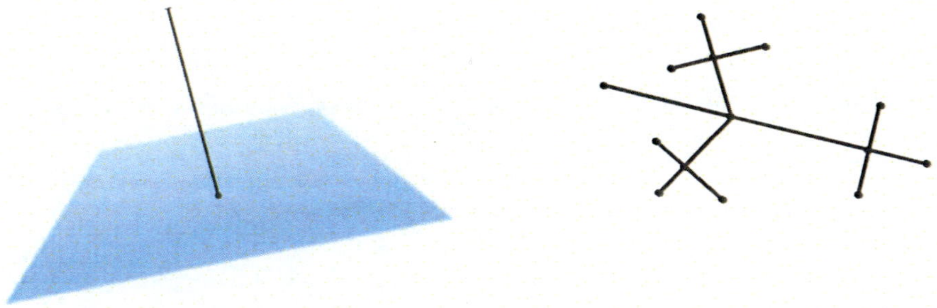

Fig. 2 These spaces are not Alexandrov spaces

$\mathbb{R}$, the half line $\mathbb{R}_+$, line segments of length greater than $\pi/\sqrt{k}$, and circles of length greater than $2\pi/\sqrt{k}$ are not Alexandrov spaces with curv $\geq k > 0$.

Remark 2.4 Since the definition of lower curvature bound is independent of the Hausdorff dimension of the length space under consideration, it is possible to omit the finiteness of the Hausdorff dimension in Definition 2.3 (see, for example, [83]). Doing so, however, introduces technical difficulties that do not arise in finite dimensions (see, for example, [47]). We refer the reader to [99, 100] for other results on infinite dimensional spaces with curvature bounded below.

The graph in Fig. 2, considered as a subset of $\mathbb{R}^2$, equipped with the length metric induced by the usual Euclidean distance, is not an Alexandrov space. Indeed, no neighborhood of a vertex of degree three or more (i.e., where three or more edges meet) has a lower curvature bound. This space, however, has curvature bounded above in the comparison sense (see, for example [12, Chap. 9]). Similarly, a subset of $\mathbb{R}^3$ consisting of a 2-plane attached to an interval by one of its endpoints with the induced length metric is not an Alexandrov space of curvature bounded below.

Alexandrov spaces have several nice topological and geometric properties. For example, they are *non-branching*, i.e., geodesics do not bifurcate. This does not hold, for instance, for the length space in Fig. 2. The Hausdorff dimension of an Alexandrov space is an integer and it equals its topological dimension.

We will denote the class of n-dimensional Alexandrov spaces with curvature bounded below by $k \in \mathbb{R}$ by $\mathrm{Alex}^n(k)$. Observe that $\mathrm{Alex}^n(k)$ contains the set of complete n-dimensional Riemannian manifolds with sec $\geq k$. This inclusion is proper, since not every Alexandrov space is homeomorphic to a topological manifold, as the examples in the following subsection illustrate. Thus, Alexandrov geometry may be seen as a generalization of Riemannian geometry and, indeed, many Riemannian theorems have corresponding analogues for Alexandrov spaces. One such result is the Bonnet–Myers theorem [12, Theorem 10.4.1], which we will use in subsequent sections.

Theorem 2.5 (Bonnet–Myers) *If X is and Alexandrov space with* curv $\geq k > 0$, *then* $\mathrm{diam}(X) \leq \pi/\sqrt{k}$. *In particular, X is compact.*

2.3 Examples and Constructions

We now list some well-known examples and constructions of Alexandrov spaces.

Complete Riemannian Manifolds with sec $\geq k$

As stated in the preceding subsection, Toponogov's comparison theorem implies that every complete Riemannian manifold with sectional curvature bounded below by k is an Alexandrov space with curv $\geq k$.

Convex Sets

Any convex subset of an Alexandrov space with curv $\geq k$ is again an Alexandrov space with curv $\geq k$.

Convex Surfaces

Every *convex surface* in $\mathbb{R}^3$, i.e., the boundary of a convex body in $\mathbb{R}^3$, equipped with the intrinsic metric induced by $\mathbb{R}^3$, is an Alexandrov space of curv ≥ 0. More generally, the boundary of any convex body in $\mathbb{R}^n$, $n \geq 3$, is an Alexandrov space of non-negative curvature. By a result of Buyalo [11, 14], any convex hypersurface N in a complete Riemannian manifold M with sectional curvature bounded below by k is an Alexandrov space of curv $\geq k$.

Gromov–Hausdorff Limits

The limit of a Gromov–Hausdorff convergent sequence of compact Alexandrov spaces with curvature bounded below by k is itself an Alexandrov space of curv $\geq k$.

Cartesian Products

Let (X, d_X) and (Y, d_Y) be Alexandrov spaces with curvature bounded below by k. **Motivated by the Pythagorean theorem, we define the** *product metric* d on the Cartesian product $X \times Y$ by letting

$$d((x_1, y_1), (x_2, y_2)) = \sqrt{d_X(x_1, x_2)^2 + d_Y(y_1, y_2)^2}$$

for all $x_1, x_2 \in X$ and $y_1, y_2 \in Y$. The space $X \times Y$ equipped with the product metric is the *direct metric product of X and Y* and is an Alexandrov space with $\operatorname{curv} \geq k$.

Cones

Let $(X, d) \in \operatorname{Alex}^n(1)$ and let $C(X) = X \times [0, \infty)/X \times \{0\}$ be the cone over X. **We define the** *Euclidean cone metric* d_C on $C(X)$ by letting

$$d_C((p, s), (q, t)) = \sqrt{s^2 + t^2 - 2st\cos(d(p, q))}$$

for all $(p, s), (q, t) \in C(X)$. Since $\operatorname{curv}(X) \geq 1$, the Bonnet–Myers theorem implies that $\operatorname{diam}(X) \leq \pi$. This ensures that d_C is indeed a metric on $C(X)$ (see, for example, [12, Propsition 3.6.13]). The metric space $(C(X), d_C)$ is the *Euclidean cone* over X and is an Alexandrov space with $\operatorname{curv} \geq 0$. In particular, the Euclidean cone over the unit round sphere $\mathbb{S}^n(1)$ is isometric to $\mathbb{R}^{n+1}$. Note that $C(X)$ contains an isometric copy of X consisting of all the points at distance one from the vertex of $C(X)$.

Suspensions

Let $(X, d) \in \operatorname{Alex}^n(1)$ and let

$$\operatorname{Susp}(X) = X \times [0, \pi]/\{X \times \{0\}, Y \times \{\pi\}\}$$

be the *suspension* of X. **Motivated by the spherical law of cosines, we define the** *spherical suspension metric* d_S on $\operatorname{Susp}(X)$ by

$$d_S((p, s), (q, t)) = \cos^{-1}(\cos(s)\cos(t) + \sin(s)\sin(t)\cos(d(p, q))$$

for all $(p, s), (q, t) \in \operatorname{Susp}(X)$. The space $(\operatorname{Susp}(X), d_S)$ is the *spherical suspension* of X and is an $(n + 1)$-dimensional Alexandrov space with $\operatorname{curv} \geq 1$. In particular, the spherical suspension of $\mathbb{S}^n(1)$ is isometric to $\mathbb{S}^{n+1}(1)$. Note that $\operatorname{Susp}(X)$ contains an isometric copy of X consisting of the set of points at distance $\pi/2$ from either vertex of the spherical suspension. In analogy with the sphere, we may think of this set as the *equator* of the suspension and of the vertices as the *poles*.

Joins

The *join* of two topological spaces X, Y is the space

$$X * Y = (X \times Y \times [0, \pi])/\sim,$$

where $\sim$ is the equivalence relation given by

$$(x, y, 0) \sim (x, y', 0) \text{ for all } x \in X \text{ and } y, y' \in Y,$$
$$(x, y, \pi) \sim (x', y, \pi) \text{ for all } x, x' \in X \text{ and } y \in Y.$$

If (X, d_X) and (Y, d_Y) are Alexandrov spaces with curv ≥ 1, then we may define a *spherical join metric* d_J on $X * Y$ so that $(X * Y, d_J)$ is an Alexandrov space with curvature bounded below by 1 and dimension $\dim X + \dim Y + 1$. We outline the definition of d_J, following [45]. Since X and Y are Alexandrov spaces with curv ≥ 1, the Euclidean cones $C(X)$ and $C(Y)$ are Alexandrov spaces with curv ≥ 0 and hence the product $C(X) \times C(Y)$ is an Alexandrov space with curv ≥ 0. Let o_X and o_Y denote, respectively, the vertices of the cones $C(X)$ and $C(Y)$. The set of points at unit distance from (o_X, o_Y) in $C(X) \times C(Y)$ is an Alexandrov space with curv ≥ 1 which can be naturally identified with the join $X * Y$. The join $X * Y$ contains isometric copies of X and Y in such a way that all points in X are at distance $\pi/2$ from all points in Y. Moreover, if $A \subset X$ and $B \subset Y$ are Alexandrov spaces isometrically embedded in X and Y, then $A * B$ is isometrically embedded in $X * Y$. We may use this observation to calculate distances between points in $X * Y$ as follows. First, note that any point in $X * Y \setminus (X \cup Y)$ has a unique coordinate representation as (x, t, y) with $x \in X$, $y \in Y$, $t \in (0, \pi/2)$. Now, given two points in the join, let A be a geodesic joining the x-coordinates and let B be a geodesic joining the y-coordinates. Note that we may think of A and B as being isometrically embedded in $S^1(1)$, since their length is at most π. Then the distance between the original points can be computed in $A * B \subset \mathbb{S}^3(1) = \mathbb{S}^1(1) * \mathbb{S}^1(1)$. Note that when Y is the two point set, $X * Y$ is isometric to the spherical suspension of X. The spherical join of two unit round spheres $\mathbb{S}^n(1)$ and $\mathbb{S}^m(1)$ is isometric to $\mathbb{S}^{n+m+1}(1)$.

Quotients

Let X be an Alexandrov space and let G be a group acting by isometries on X with closed orbits. The orbit space X^* has a metric given by

$$d_Q(p^*, q^*) = \inf\{d(x, y) : x \in G(p), \ y \in G(q)\}$$

for all $p^*, q^* \in X^*$. If $X \in \text{Alex}^n(k)$, then X^* is also an Alexandrov space with curv $\geq k$. This is a consequence of the fact that the orbit projection map $\mathsf{pr}\colon X \to X^*$ is a *submetry*, i.e., the image under pr of a metric ball of radius $r > 0$ and center $p \in X$ is a metric ball in X^* with radius r and center $p^* \in X^*$.

Doubles and Glued Spaces

Let $X_1, X_2 \in \mathrm{Alex}^n(k)$ with non-empty boundary (for the definition of boundary see Sect. 2.4 below) and let $f : \partial X_1 \to \partial X_2$ be an isometry. The space $X = X_1 \cup X_2/(p \sim f(p))$ has a metric with curv $\geq k$ given by

$$d(p_1, p_2) = \inf\{\, d_1(p_1, q) + d_2(f(q), p_2) \;:\; q \in \partial X_1 \,\}.$$

This was first proved for doubles of Alexandrov spaces with boundary by Perelman [77]. The general gluing result was obtained by Petrunin [81]. It follows from these results that the *double disc*, i.e., the gluing of two copies of a disc in $\mathbb{R}^2$ along their isometric boundaries, is an Alexandrov space with non-negative curvature whose underlying topological space is homeomorphic to a 2-sphere.

We may use the preceding constructions to generate Alexandrov spaces that are neither topological manifolds nor orbifolds. Consider, for example, the complex projective plane $\mathbb{C}P^2$. Equipped with its canonical Fubini–Study metric, $\mathbb{C}P^2$ is a Riemannian manifold with sec ≥ 1 and is, therefore, an Alexandrov space with curv ≥ 1. Hence the spherical suspension $\mathrm{Susp}(\mathbb{C}P^2)$ is again an Alexandrov space with curv ≥ 1 and is homeomorphic neither to a topological manifold nor to an orbifold. Further examples of Alexandrov spaces with curvature bounded below include certain warped products [2, 4] and stratified spaces [8].

2.4 Local Structure

The local structure of an Alexandrov space is determined by the *space of directions*. To define this space, we first define angles between geodesics. The *angle* between two geodesics $[pq]$, $[pr]$ in an Alexandrov space X is defined as

$$\angle qpr = \lim_{q_1, r_1 \to p} \{\tilde{\angle} q_1 \overline{p r}_1 \,:\, q_1 \in [pq],\; r_1 \in [pr]\}.$$

Geodesics that make an angle zero determine an equivalence class called *tangent direction*. The set of tangent directions at a point $p \in X$ is denoted by Σ'_p and, when equipped with the angle distance $\angle$, the set Σ'_p is a metric space. This space may not be complete, however, as one can see by considering directions at a point in the boundary of a unit disc in the Euclidean plane. The completion of $\left(\Sigma'_p, \angle\right)$ is called the *space of directions of X at p* and is denoted by Σ_p. It corresponds, in Alexandrov geometry, to the unit tangent sphere in Riemannian geometry. The space of directions Σ_p is an Alexandrov space with curv ≥ 1 and dimension $\dim X - 1$. Moreover, Σ_p is isometric to the unit round sphere $\mathbb{S}^{n-1}(1)$ on a dense set $\mathcal{R}_X$, called the set of *(metrically) regular points*. The set $\mathcal{S}_X = X \setminus \mathcal{R}_X$ is the set of *(metrically) singular points*. The following example shows that the set $\mathcal{S}_X$ may be dense in X.

Example 2.6 (Otsu–Shioya [75, p. 632]) Let P be a convex polyhedron in $\mathbb{R}^3$. For any vertex $p \in P$, let $\angle(P, p)$ be the sum of all inner angles at p of faces of P having p as a vertex. Since P (equipped with the length metric induced by $\mathbb{R}^3$) is an Alexandrov space of non-negative curvature, $\angle(P, p) \le 2\pi$ and the space of directions Σ_p is a circle of length $\angle(P, p)$. Thus, a vertex $p \in P$ is a metrically singular point if and only if $\angle(P, p) < 2\pi$. We now define inductively a Hausdorff-convergent sequence $\{X_k\}_{k=1}^{\infty}$ of convex polyhedra in $\mathbb{R}^3$ with limit X. Since Hausdorff convergence implies Gromov–Hausdorff convergence, X will be an Alexandrov space of non-negative curvature. We define the X_i so that the set of metrically singular points in X is dense. Let X_1 be a regular tetrahedron in $\mathbb{R}^3$ whose barycenter is the origin $o \in \mathbb{R}^3$. Assume that X_k for some $k > 1$ has been defined. Let us define X_{k+1}. Let $\{\varepsilon_i\}_{i=1}^{\infty}$ be a monotone decreasing sequence of positive numbers converging to zero. Assume that $0 < \varepsilon_i < 1$ for all i and let $\varepsilon = \Pi_{j=1}^{\infty}(1 - \varepsilon_j)$; note that, by construction, $\varepsilon > 0$. Take the barycentric subdivision of X_k and push all the new vertices outward slightly along rays emanating from o while keeping the original vertices of X_k to obtain the convex tetrahedron X_{k+1}. We may assume that

$$2\pi - \angle(X_{k+1}, p) \ge (1 - \varepsilon_k)(2\pi - \angle(X_k, p))$$

for any vertex p of X_k. Define $X \subset \mathbb{R}^3$ to be the Hausdorff limit of $\{X_k\}$. Then X is a non-negatively curved Alexandrov space. For any k and any vertex p of X_k, we obtain

$$\begin{aligned}\lim_{i\to\infty}(2\pi - \angle(X_i, p)) &\ge \Pi_{i=1}^{\infty}(1 - \varepsilon_{k+1})(2\pi - \angle(X_k, p))\\ &\ge \varepsilon(2\pi - \angle(X_k, p))\\ &> 0.\end{aligned}$$

The length of the space of directions of X at p is $\lim_{i\to\infty} \angle(X_i, p) < 2\pi$. Thus any vertex of X_k for any k is a singular point of X. Since the maximal length of all the edges of X_k tends to zero as $k \to \infty$, the set S_X of singular points is dense in X.

Example 2.7 Let M be a complete Riemannian manifold with sectional curvature bounded below by k. Recall that, if G is a compact Lie group acting effectively and isometrically on M, then the orbit space M^* is an Alexandrov space with curv $\ge k$. In this case the space of directions Σ_{p^*} at $p^* \in M^*$ consists of geodesic tangent directions and is isometric to $\mathbb{S}_p^{\perp}/G_p$, where $\mathbb{S}_p^{\perp}$ is the unit normal sphere to the orbit $G(p)$ at $p \in M$.

Definition 2.8 Let X be an Alexandrov space and fix $p \in X$. The *tangent cone* of X at a p is the Euclidean cone over the space of directions of X at p. We will denote it by T_pX.

By construction T_pX is an Alexandrov space with curv ≥ 0 and $\dim T_pX = \dim X$. Note that the tangent cone of a complete Riemannian manifold at any one of its points is the usual tangent space and is isometric to a Euclidean space.

The local structure of Alexandrov spaces is given by the following theorem (see [77]).

Theorem 2.9 (Conical Neighborhood Theorem (Perelman)) *If X is an Alexandrov space, then every $p \in X$ has a neighborhood pointed-homeomorphic to T_pX.*

Remark 2.10 It is conjectured that the homeomorphism in the preceding theorem should be bi-Lipschitz.

The conical neighborhood theorem implies that the local topology of an Alexandrov space X at a point p is determined by the space of directions Σ_p. Thus, since Σ_p is an Alexandrov space with curvature bounded below by 1, it is important to determine the possible homeomorphism types of such Alexandrov spaces. In dimensions 2 and 3 this classification is complete (see the next section). In dimensions $n \geq 4$, however, the classification problem is open in full generality, even in the case of Riemannian manifolds (see [101] and references therein).

Having defined the space of directions at a point, we may now define the boundary of an Alexandrov space X. If $\dim X = 1$, then X is a manifold (possibly with boundary). We define the boundary of an Alexandrov space inductively by letting

$$\partial X = \{p \in X : \partial \Sigma_p \neq \varnothing\}$$

if $\dim X > 1$. The boundary of X is a closed subset of codimension 1. Note that if $\dim X \leq 2$, then X is a topological manifold (possibly with boundary).

An Alexandrov space is an *Alexandrov manifold* if it is homeomorphic to a topological manifold. We will say that an Alexandrov space is *topologically regular* if every space of directions is homeomorphic to a sphere. Clearly, a topologically regular Alexandrov space is an Alexandrov manifold, but the converse is not necessarily true, as the following example shows.

Example 2.11 Recall that the *Poincaré homology sphere*, which we denote by P^3, is diffeomorphic to a quotient of the 3-sphere by a free smooth action of the binary icosahedral group I^*. Thus $P^3 = \mathbb{S}^3/I^*$ is a compact non-simply-connected 3-manifold without boundary and with the same integral homology groups as $\mathbb{S}^3$. We may assume that I^* acts orthogonally on the unit round 3-sphere, which implies that P^3 inherits a Riemannian metric with constant sectional curvature one. In particular, P^3 is an Alexandrov space of curv ≥ 1 and, hence, its the double spherical suspension $(\mathrm{Susp}^2(P^3), d)$ is a 5-dimensional Alexandrov space with curv ≥ 1. By the Double Suspension Theorem of Edwards and Cannon, $\mathrm{Susp}^2(P)$ is homeomorphic to $\mathbb{S}^5$ (see [15, 23, 24]). It follows that $(\mathrm{Susp}^2(P), d)$ is a five-dimensional Alexandrov manifold. On the other hand, $(\mathrm{Susp}^2(P), d)$ is not topologically regular, since the spaces of directions at the poles of the double suspension are homeomorphic to $\mathrm{Susp}(P)$, which is not a manifold.

Definition 2.12 Let $X \in \mathrm{Alex}^n(k)$. A subset $E \subset X$ is *extremal* if, for $p \in X$ and $q \in E$ with $d(p,q) = d(p,E)$, one has $\boldsymbol{\Sigma}_q = B(p', \frac{\pi}{2})$, where $p' \in \boldsymbol{\Sigma}_q$ is the direction of a geodesic from q to p.

A point p in an Alexandrov space X is extremal if and only if $\mathrm{diam}(\boldsymbol{\Sigma}_p) \leq \pi/2$. Observe that ∂X is extremal. By work of Perelman and Petrunin [78], each extremal set $E \subset X$ can be decomposed into a disjoint union of topological manifolds; along with an n-dimensional open set, this determines a stratification of X into topological manifolds.

2.5 A Riemannian Digression

Alexandrov spaces play an important role in Riemannian geometry and arise, for example, in the following context. Recall that, equipped with the Gromov–Hausdorff distance d_{GH} defined in Sect. 2.1, the collection $\mathcal{X}$ of isometry classes of compact metric spaces is itself a metric space. A key fact is that the class of (isometry classes of) compact Riemannian n-manifolds

$$\mathcal{M}_k^D(n) = \left\{ (M^n, g) \ : \ \sec(M^n) \geq k, \mathrm{diam}(M^n) \leq D \right\} \subset (\mathcal{X}, d_{\mathrm{GH}})$$

is precompact under the Gromov–Hausdorff topology for all $n \in \mathbb{N}$, $k \in \mathbb{R}$ and $D > 0$ (see [12, Chap. 7]. Moreover, the points in the closure of $\mathcal{M}_k^D(n)$ are Alexandrov spaces:

Theorem 2.13 (Burago et al. [13]; Grove and Petersen [44]) *The closure* $\overline{\mathcal{M}_k^D}(n)$ *consists of Alexandrov spaces with* $\mathrm{curv} \geq k$, $\mathrm{diam} \leq D$, *and* $\dim \leq n$.

Alexandrov spaces therefore naturally appear as limit spaces of Riemannian manifolds. It is not known, however, if every compact Alexandrov space is the limit of a sequence of compact Riemannian manifolds with sectional curvature uniformly bounded below. In the case where the sequence is *non-collapsed*, i.e., where the limit space has the same dimension as the elements of the sequence, it follows from Perelman's stability theorem that the limit space must be a topological manifold (see [57, 79]).

Theorem 2.14 (Stability Theorem (Perelman)) *Let X, Y be compact n-dimensional Alexandrov spaces of* $\mathrm{curv} \geq k$. *Then there exists an $\varepsilon = \varepsilon(X) > 0$ such that, if $d_{GH}(X,Y) < \varepsilon$, then Y is homeomorphic to X.*

This theorem, in combination with Theorem 2.13, implies the following well-known finiteness result in Riemannian geometry (see, for example, [42]).

Theorem 2.15 (Riemannian Homeomorphism Finiteness Theorem) *For each $n \in N$, $k \in \mathbb{R}$, and $D, v > 0$, the class $\mathcal{M}_{k,v}^{D}(n)$ of compact Riemannian n-manifolds with* diam $\leq D$, sec $\geq k$, *and volume* vol $\geq v$, *contains at most finitely many homeomorphism types.*

Note that a similar finiteness result holds for Alexandrov spaces. Although the class of compact Alexandrov spaces with lower curvature bound k is closed in the Gromov–Hausdorff topology, collapse may occur, that is, it is possible for a Gromov–Hausdorff converging sequence of spaces in $\mathrm{Alex}^n(k)$ to have a limit of dimension less than n. We may observe this phenomenon by considering, for example, a flat torus T^n and rescaling its Riemannian metric by $1/i$ with $i = 1, 2, \ldots$. In this way we get a sequence of flat n-dimensional tori T_i^n whose diameter converges to zero as $i \to \infty$. In this case, the sequence $\{T_i^n\}_{i=1}^{\infty}$ collapses to a point. Collapse imposes strong restrictions on the structure of Riemannian manifolds; for more information we refer the reader to [42, 87]. We will discuss the topology of collapsed 3-dimensional Alexandrov spaces in Sect. 7.

3 Three-Dimensional Alexandrov Spaces

In this section we discuss the basic topology and geometry of closed (i.e., compact and without boundary) three-dimensional Alexandrov spaces (or, for short, closed *Alexandrov* 3-*spaces*). We will focus our attention on those that are not homeomorphic to 3-manifolds, as this is where new phenomena arise. The symbol "$\approx$" will denote homeomorphism between topological spaces.

3.1 Basic Structure

Let us first consider the topological structure of closed Alexandrov 3-spaces, following [29]. We refer the reader to [51] for basic results in 3-manifold topology.

Recall that a closed Alexandrov space of dimension one must be homeomorphic to a circle. Then, by Perelman's conical neighborhood theorem, a 2-dimensional Alexandrov space must be homeomorphic to a topological manifold, possibly with boundary. The topological classification of closed, positively curved Alexandrov spaces of dimension two follows now from the Bonnet–Myers theorem, which implies that the fundamental group of a closed, positively curved Alexandrov space must be finite. Therefore, any closed two-dimensional Alexandrov space with curv ≥ 1 is homeomorphic to $\mathbb{S}^2$ or to $\mathbb{R}P^2$. It follows that $\mathbb{S}^2$ and $\mathbb{R}P^2$ are the only possible spaces of directions of an Alexandrov 3-space without boundary. Hence, by the conical neighborhood theorem, an Alexandrov 3-space X without boundary is a topological 3-manifold if and only if each one of its points has space of directions homeomorphic to $\mathbb{S}^2$ i.e., if X is topologically regular).

Let X be a closed Alexandrov 3-space and assume that X is not a topological manifold. Hence at least one point in X has space of directions homeomorphic to $\mathbb{R}P^2$. Since X is compact, the conical neighborhood theorem implies that there are finitely many points in X whose space of directions is $\mathbb{R}P^2$. After removing from X sufficiently small open neighborhoods of these topologically singular points we get a compact non-orientable 3-manifold X_o with a finite number of $\mathbb{R}P^2$-boundary components where we glue in cones over $\mathbb{R}P^2$. It is not difficult to see that X_o must have an even number of boundary components (cf. [51, Proof of Theorem 9.5]) or, equivalently, that X must have an even number of topologically singular points. Let $D(X_o)$ be the double of X_o and consider the natural decomposition of $D(X_o)$ as the union of two copies of X_o glued along ∂X_o. From the Mayer–Vietoris sequence for this decomposition of $D(X_o)$ we obtain that

$$\chi(D(X_o)) = 2\chi(X_o) - \chi(\partial X_o). \tag{1}$$

Since $D(X_0)$ is a closed 3-manifold, its Euler characteristic is zero. Hence, Eq. (1) implies that $\chi(\partial X_o)$ is even. Since each connected component of ∂X_o is a real projective space and $\chi(\mathbb{R}P^2) = 1$, it follows that X_o has an even number of boundary components. Therefore, X has an even number of topologically singular points.

Example 3.1 The real projective plane $\mathbb{R}P^2$, equipped with its canonical Riemannian metric of constant sectional curvature 1, is a closed 2-dimensional Alexandrov space with curv ≥ 1. Thus, its spherical suspension $\mathrm{Susp}(\mathbb{R}P^2)$ is a closed Alexandrov 3-space with curv ≥ 1 with exactly two topologically singular points, namely, the poles of the suspension. In this case, $X_o \approx \mathbb{R}P^2 \times [0, 1]$ and we obtain $\mathrm{Susp}(\mathbb{R}P^2)$ after capping off each boundary component of X_o with a cone over $\mathbb{R}P^2$. We may also obtain the spherical suspension $\mathrm{Susp}(\mathbb{R}P^2)$ as a quotient of the unit round 3-sphere $\mathbb{S}^3(1)$ as follows. Recall first that $\mathbb{S}^3(1)$ is isometric to the spherical suspension of the unit round 2-sphere $\mathbb{S}^2(1)$. Consider now the involution $\iota\colon \mathbb{S}^3(1) \to \mathbb{S}^3(1)$ corresponding to the suspension of the antipodal map on $\mathbb{S}^2(1)$. This involution is an orientation-reversing isometry of $\mathbb{S}^3(1)$ and its metric quotient $\mathbb{S}^3(1)/\iota$ is isometric to the spherical suspension of $\mathbb{R}P^2$. Note that the involution $\iota\colon \mathbb{S}^3(1) \to \mathbb{S}^3(1)$ has exactly two isolated fixed points; these fixed points project down to the poles of $\mathrm{Susp}(\mathbb{R}P^2)$, giving rise to the two topologically singular points of this space.

The preceding example illustrates a general situation. Given a topologically singular Alexandrov 3-space X, there is a closed, orientable 3-manifold Y and an orientation reversing involution $\iota\colon Y \to Y$ with only isolated fixed points such that $X \approx Y/\iota$. It is important to note that $\iota\colon Y \to Y$ is conjugate to a smooth involution on Y. Hence X is homeomorphic to a smooth non-orientable 3-orbifold. The preceding properties imply that X is the base of a two-fold branched cover $\mathrm{pr}\colon Y \to X$ whose total space Y is a closed, orientable 3-manifold and whose branching set is the set of points with space of directions homeomorphic to $\mathbb{R}P^2$. In sum, up to homeomorphism, any closed Alexandrov 3-space is either a 3-

manifold or a quotient of a closed orientable 3-manifold by and orientation reversing smooth involution with only isolated fixed points. Note that the latter spaces are homeomorphic to the *singular* 3-*manifolds* (without boundary) introduced by Quinn in [85] (see [54, 55] as well as [64, Open Problem 6], which asks to develop a theory for such spaces).

The construction of the orientable double branched cover $\mathsf{pr}\colon Y \to X$ relies only on the topology of X. The following theorem brings the geometry of X into play, allowing us to lift the metric of X to Y and turning the double branched cover $\mathsf{pr}\colon Y \to X$ into a metric object compatible with the geometry of X, as illustrated in Example 3.1.

Theorem 3.2 ([29]) *Let $X \in \mathrm{Alex}^3(k)$ be topologically singular and let Y be the orientable double branched cover of X. Then the following hold:*

1. *The metric in X can be lifted to Y, so that $Y \in \mathrm{Alex}^3(k)$.*
2. *The involution $\iota\colon Y \to Y$ is an isometry.*
3. *The space of directions $\Sigma_{p'}Y \approx \mathbb{S}^2$ at a fixed point p' of the involution $\iota\colon Y \to Y$ is the canonical Alexandrov double cover of $\Sigma_{p'}X \approx \mathbb{R}P^2$.*

A detailed proof of the preceding theorem would take us beyond the introductory treatment of Alexandrov spaces in this survey. Thus we will only discuss the main ideas in the proof; we refer the reader to [20, Sect. 2.1] for more details (cf. [46, Sects. 2 and 5] and [50]). The initial point in the proof of Theorem 3.2 is the observation that the set $X_o \subset X$ of topologically regular points of X is convex in X. Letting $d_{X_0} = d_X|_{X_0}$ be the restriction of the metric on X to X_o, the convexity of X_o implies that the space (X_o, d_{X_o}) is a non-complete length space that is also a *k-domain*, i.e., the comparison property T_k holds for any geodesic triangle in (X_o, d_{X_o}). Since X_o is a non-orientable topological 3-manifold, it has an orientable double cover Y_o and we may lift the metric d_{X_o} to a metric d_{Y_o} on Y_o. By construction, the metric space (Y_o, d_{Y_o}) is a length space locally isometric to (X_o, d_o) (see [12, Chaps. 2.2 and 3.4]). Thus (Y_o, d_{Y_o}) has curvature locally bounded below by k and its metric completion is homeomorphic to the two-fold branched cover Y of X. One then shows, using work of Li [60], that (Y, d_Y) also has curvature bounded below by k. This then implies that the involution $\iota\colon Y \to Y$ is an isometry. Along the way one constructs the space of directions for Y at a fixed point of the involution $\iota\colon Y \to Y$, showing part (3) of the Theorem. We refer the reader to [38] for further explicit examples, besides Example 3.1, of spaces arising from orientation-reversing involutions on closed orientable 3-manifolds.

4 Spaces with Positive or Non-negative Curvature

We now turn our attention to the topological classification of Alexandrov 3-spaces with positive or non-negative curvature. In the Riemannian category this classification follows from Hamilton's classification of closed 3-manifolds with positive or non-negative Ricci curvature [48, 49].

Recall that positively curved Alexandrov spaces arise as spaces of directions and determine the local topology of Alexandrov spaces via the conical neighborhood theorem (see Sect. 2.4). Thus the classification of positively curved spaces is of fundamental importance. This is, however, a challenging problem which has only been solved in dimensions 2 and 3, even in the Riemannian case. We refer the reader to [43, 101] for more information on positively curved Riemannian manifolds.

Recall that, by the Bonnet–Myers theorem, closed Alexandrov spaces with positive curvature have finite fundamental group. Thus, as we have previously seen, a closed 2-dimensional Alexandrov space of positive curvature is homeomorphic to the 2-sphere or to the real projective plane.

The topological type of Alexandrov 3-spaces with positive curvature is given by the following theorem (see [29, 50]). Our presentation follows [29]. Recall that a *spherical* 3-*manifold* is a 3-manifold homeomorphic to $\mathbb{S}^3/\Gamma$, where Γ is a finite group acting freely and orthogonally on the 3-sphere. Note that every spherical 3-manifold is homeomorphic to a three-dimensional *spherical space form*, i.e., a closed Riemannian 3-manifold with constant positive sectional curvature. Conversely, every three-dimensional spherical space form is a spherical 3-manifold.

Theorem 4.1 (Alexandrov 3-Spaces of Positive Curvature) *A closed Alexandrov 3-space of positive curvature is homeomorphic to a spherical 3-manifold or to* $\mathrm{Susp}(\mathbb{R}P^2)$.

Proof Let X be a closed Alexandrov 3-space with positive curvature. We may assume, after re-scaling the metric if necessary, that $\mathrm{curv}\, X \geq 1$. Suppose first that X is a manifold. Then, by the Bonnet–Myers theorem, X has finite fundamental group. It follows then from Perelman's proof of the Poincaré Conjecture and Thurston's Elliptization Conjecture that X must be homeomorphic to a spherical manifold, including the 3-sphere.

Suppose now that X is not a topological manifold and let X' be the set of points in X whose space of directions is homeomorphic to $\mathbb{R}P^2$. By hypothesis, X' is nonempty and, by the conical neighborhood theorem, each point in X' has a neighborhood homeomorphic to the Euclidean cone $C_0(\mathbb{R}P^2)$. Since X is compact, the set X' is finite.

Let $p_1, \ldots, p_k$ be the points in X'. After removing a neighborhood of each p_i homeomorphic to $C_0(\mathbb{R}P^2)$ we obtain a topological 3-manifold X_o whose boundary consists of k copies of $\mathbb{R}P^2$.

Let $\mathsf{pr}\colon Y \to X$ be the two-fold branched cover over X with branching set X'. Let $q_i = \mathsf{pr}^{-1}(p_i)$, $i = 1, \ldots, k$, and let $Y' = \{q_1, \ldots, q_k\}$. By Theorem 3.2, Y is an Alexandrov space with $\mathrm{curv} \geq 1$. Hence, by Theorem 2.5, Y has finite fundamental group. On the other hand, $\pi_1(Y) \simeq \pi_1(Y \setminus Y')$, since Y' is a finite set of points in Y. Since $\mathsf{pr}\colon Y \setminus Y' \to X \setminus X'$ is a regular two-fold cover, $\mathsf{pr}_*(\pi_1(Y \setminus Y'))$ is a subgroup of index 2 in $\pi_1(X \setminus X')$. Hence, $\pi_1(X \setminus X')$ is finite. It follows from Epstein's theorem (see [51, Chap. 9]) and Perelman's proof of the Poincaré Conjecture, that $X \setminus X'$ is homeomorphic to $\mathbb{R}P^2 \times [0, 1]$. Thus, $k = 2$ and we conclude that X is homeomorphic to $\mathrm{Susp}(\mathbb{R}P^2)$, as desired. Observe that Y is homeomorphic to $\mathbb{S}^3$ and, by work of Hirsch, Smale [52] and Livesay [61], the action of $\mathbb{Z}_2$ corresponding to the two-fold branched cover is equivalent to a linear action given by the suspension of the antipodal map on $\mathbb{S}^2$ (cf. Example 3.1). □

Corollary 4.2 *A closed, simply-connected three-dimensional Alexandrov space of positive curvature is homeomorphic to* $\mathbb{S}^3$ *or to* $\mathrm{Susp}(\mathbb{R}P^2)$.

Corollary 4.3 *The space of directions of a* 4*-dimensional Alexandrov space without boundary is homeomorphic to* $\mathrm{Susp}(\mathbb{R}P^2)$ *or to a spherical* 3*-manifold.*

Corollary 4.4 *A closed* 4*-dimensional Alexandrov space of curvature bounded below by* 1 *and diameter greater than* $\pi/2$ *is homeomorphic to the suspension of a spherical* 3*-manifold or to* $\mathrm{Susp}^2(\mathbb{R}P^2)$*, the double suspension of* $\mathbb{R}P^2$.

Corollary 4.2 follows from Perelman's proof of the Poincaré Conjecture. Corollary 4.3 follows from the fact that the space of directions at any point of an n-dimensional Alexandrov space is isometric to a compact $(n-1)$-dimensional Alexandrov space with curvature bounded below by 1. Note that Corollary 4.3 implies that 4-dimensional Alexandrov spaces without boundary are, locally, orbifolds without boundary. Finally, Corollary 4.4 follows from the fact that an n-dimensional Alexandrov space of curvature bounded below by 1 and diameter greater than $\pi/2$ is homeomorphic to the suspension of a compact $(n-1)$-dimensional Alexandrov space of curvature bounded below by 1.

Corollary 4.5 *Let* X^n *be an* n*-dimensional Alexandrov manifold. If* $n \leq 4$*, then* X^n *is topologically regular.*

Proof If $n \leq 3$, the conclusion follows from the fact, recalled at the end of Sect. 2, that every 1- or 2-dimensional Alexandrov space must be homeomorphic to a topological manifold. Suppose now that $n = 4$ and let X^4 be an Alexandrov 4-manifold. By the conical neighborhood theorem, any sufficiently small neighborhood U of a point $p \in X^4$ is homeomorphic to the cone over the space of directions Σ_p at p. Since a cone over a non-simply-connected 3-manifold cannot be homeomorphic to the 4-ball D^4, the only case we need to consider is when $\Sigma_p X$ is homeomorphic to $\mathrm{Susp}(\mathbb{R}P^2)$. In this case, a simple calculation using the long exact sequence in homology of the pair $(U, U - p)$ implies that some homology group $H_k(U, U - p)$ is not isomorphic to $H_k(D^4, \mathbb{S}^3)$. Thus X cannot be a topological manifold. □

Corollary 4.5 is optimal, since, by Example 2.11, Alexandrov n-manifolds, $n \geq 5$, are not necessarily topologically regular.

The ideas in the proof of Theorem 4.1 can also be used to provide a complete description of closed Alexandrov 3-spaces of non-negative curvature. We denote the non-orientable $\mathbb{S}^2$-bundle over $\mathbb{S}^1$ by $\mathbb{S}^2 \tilde{\times} \mathbb{S}^1$, and the suspension of $\mathbb{R}P^2$ by $\mathrm{Susp}(\mathbb{R}P^2)$. Given two Alexandrov 3-spaces X, Y, we denote their connected sum by $X\#Y$, i.e., $X\#Y$ is the space obtained by removing an open 3-ball from X, an open 3-ball from Y, and then identifying the boundaries of the resulting topological spaces.

Theorem 4.6 (Alexandrov 3-Spaces of Non-negative Curvature [29]; cf. [20]) *Let X^3 be a closed, non-negatively curved Alexandrov 3-space.*

(1) *If X^3 is a topological manifold, then one of the following holds:*
- X^3 *is homeomorphic to a spherical space form,*
- X^3 *is homeomorphic to* $\mathbb{S}^2 \times \mathbb{S}^1$, $\mathbb{R}P^2 \times \mathbb{S}^1$, $\mathbb{R}P^3 \# \mathbb{R}P^3$ *or* $\mathbb{S}^2 \tilde{\times} \mathbb{S}^1$*; or*
- X^3 *is isometric to a closed, flat three-dimensional space form.*

(2) *If X^3 has a point with space of directions homeomorphic to $\mathbb{R}P^2$, then either:*
- X^3 *is homeomorphic to* $\mathrm{Susp}(\mathbb{R}P^2)$, $\mathrm{Susp}(\mathbb{R}P^2) \# \mathrm{Susp}(\mathbb{R}P^2)$, $\mathbb{R}P^3 \# \mathrm{Susp}(\mathbb{R}P^2)$ *or*
- X^3 *is isometric to a quotient of a closed, orientable, flat three-dimensional manifold by an orientation reversing isometric involution with only isolated fixed points.*

Let us briefly discuss the proof of this theorem. Let X be a closed Alexandrov 3-space with curv ≥ 0. As in the proof of Theorem 4.1, we consider two possibilities, depending on whether or not X is a topological manifold.

Suppose first that X is a topological manifold. We have two possibilities: either the fundamental group $\pi_1(X)$ is finite or not. If $\pi_1(X)$ is finite, then, as in the proof of Theorem 4.1, X is homeomorphic to a spherical space form, by Perelman's resolution of Thurston's elliptization conjecture.

Suppose now that $\pi_1(X)$ is infinite. Then, the splitting theorem for non-negatively curved Alexandrov spaces [12, Theorem 10.5.1] implies that $\tilde{X}$, the universal cover of X, is isometric to a product $\mathbb{R} \times \tilde{Y}$, where $\tilde{Y}$ is a simply-connected Alexandrov 2-space with curv ≥ 0.

Kwun and Tollefson [59], and Luft and Sjerve [63], classified the involutions with only isolated fixed points on closed, orientable, flat three-dimensional space forms and their orbit spaces have been classified. These orbit spaces are the spaces in the second item of part (2) of Theorem 4.6 above.

4.1 Spaces with Positive or Non-negative Ricci Curvature

One can generalize Theorems 4.1 and 4.6 to closed Alexandrov 3-spaces with an arbitrary lower curvature bound and with positive or non-negative Ricci curvature in the sense of Lott–Sturm–Villani (see [62, 95, 96]). In this case one obtains the same list of spaces as in the case of positive or non-negative curvature in the triangle comparison sense (see [20]).

Theorem 4.7 (Alexandrov 3-Spaces of Positive Ricci Curvature [20]) *Let X^3 be a closed* $\mathsf{CD}^*(2,3)$*-Alexandrov space.*

(1) *If X^3 is a topological manifold, then it is homeomorphic to a spherical space form.*
(2) *If X^3 is not a topological manifold, then it is homeomorphic to* $\mathrm{Susp}(\mathbb{R}P^2)$.

Theorem 4.8 (Alexandrov 3-Spaces of Non-Negative Ricci Curvature [20]) *Let X^3 be a closed* $\mathsf{CD}^*(0,3)$*-Alexandrov space.*

(1) *If X^3 is a topological manifold, then one of the following holds:*
   - *X^3 is homeomorphic to a spherical space form,*
   - *X^3 is homeomorphic to $\mathbb{S}^2 \times \mathbb{S}^1$, $\mathbb{R}P^2 \times \mathbb{S}^1$, $\mathbb{R}P^3 \# \mathbb{R}P^3$ or $\mathbb{S}^2 \tilde{\times} \mathbb{S}^1$; or*
   - *X^3 is isometric to a closed, flat three-dimensional space form.*

(2) *If X^3 is not a topological manifold, then*
   - *X^3 is homeomorphic to* $\mathrm{Susp}(\mathbb{R}P^2)$, $\mathrm{Susp}(\mathbb{R}P^2) \# \mathrm{Susp}(\mathbb{R}P^2)$, $\mathbb{R}P^3 \# \mathrm{Susp}(\mathbb{R}P^2)$ *or*
   - *X^3 is isometric to a quotient of a closed, orientable, flat three- dimensional manifold by an orientation reversing isometric involution with only isolated fixed points.*

Observe that, since X is closed, then its Hausdorff measure $\mathcal{H}^3(X)$ is finite. Hence, the equivalence of the CD and CD^* conditions for (essentially non-branching) spaces with finite measures due to Cavalletti–Milman (see [17, Corollary 13.7]) implies that Theorems 4.7 and 4.8 are still valid for $\mathsf{CD}(2,3)$- and $\mathsf{CD}(0,3)$-Alexandrov spaces, respectively.

5 Topological Results

5.1 Geometrization

In view of Perelman's resolution of Thurston's geometrization conjecture, geometric 3-manifolds can be considered the building blocks of arbitrary 3-dimensional closed manifolds. It is then natural to ask about the corresponding notion for Alexandrov spaces. Recall that the eight Thurston geometries are $\mathbb{S}^3$, $\mathbb{E}^3$, $\mathbb{H}^3$, $\mathbb{S}^2 \times \mathbb{R}$, $\mathbb{H} \times \mathbb{R}$, Nil, Sol and $\widetilde{\mathrm{SL}_2(\mathbb{R})}$ (see [88]). A closed Alexandrov 3-space X^3 is *geometric* if it can be written as a quotient of one of the eight Thurston geometries by some cocompact lattice. We say that X^3 admits a *geometric decomposition* if there exists a collection of spheres, projective planes, tori and Klein bottles that decompose X^3 into geometric pieces. The following result is proved using the existence of the double branched cover, outlined in the preceding section, in combination with Dinkelbach and Leeb's work on equivariant Ricci flow [21].

Theorem 5.1 (Geometrization of Alexandrov 3-spaces [29]) *A closed Alexandrov 3-space admits a geometric decomposition into geometric Alexandrov 3-spaces.*

The proof of this theorem relies in a key way on the Ricci flow for smooth Riemannian metrics on 3-manifolds. It would be interesting to determine whether it is possible to define Ricci flow for general Alexandrov 3-spaces with no regularity assumptions on the metric.

5.2 *Simply-Connected Spaces*

In the sense of topology, the simplest spaces that one can consider are *contractible* spaces in which, a fortiori, all homotopy groups vanish. However, it is easy to show that, as in the manifold case, no closed n-dimensional Alexandrov space X, $n \geq 2$, is contractible. Indeed, on the one hand, if X is orientable, then the top homology group $H_n(X; \mathbb{Z}) \cong \mathbb{Z}$ (see, for example, [65, Theorem 1.8]). Then, the Hurewicz theorem readily implies that $\pi_n(X) \neq 0$. If, on the other hand, we assume that X is non-orientable, then [65, Corollary 5.7] gives that $H_{n-1}(X; \mathbb{Z}) \neq 0$, and another application of the Hurewicz theorem grants again that X cannot be contractible. Thus, if one aims at a deeper understanding of the topology of Alexandrov 3-spaces, a natural step is to consider other simple (topologically speaking) classes of spaces, such as simply-connected spaces.

By Perelman's proof of the Poincaré conjecture, a closed, simply-connected 3-manifold must be homeomorphic to the 3-sphere. By Poincaré duality and the Hurewicz theorem, a closed 3-manifold is simply-connected if and only if it is a homotopy sphere. This is no longer the case for Alexandrov 3-spaces, as one sees by considering Susp($\mathbb{R}P^2$). One can still show, however, that a closed Alexandrov 3-space that is also a homotopy sphere is homeomorphic to $\mathbb{S}^3$, thus obtaining an analogue for Alexandrov 3-spaces of the Generalized Poincaré Conjecture (see [29, Proposition 1.4]). On the other hand, there exist closed, geometric, simply-connected Alexandrov 3-spaces that are not homeomorphic to the 3-sphere if and only if the corresponding Thurston geometry is not one of Nil, $\widetilde{\mathrm{SL}_2(\mathbb{R})}$ or Sol. To rule out the Nil, $\widetilde{\mathrm{SL}_2(\mathbb{R})}$ and Sol geometries, one proves that there are no orientation-reversing involutions with only isolated fixed points on closed geometric 3-manifolds with one of these geometries.

A complete topological description of closed, simply-connected three-dimensional Alexandrov spaces seems currently beyond reach. A naive conjecture would be that any such space is the connected sum of $\mathbb{S}^3$ and suspensions over projective planes. This is true under the presence of an action of a compact Lie group of positive dimension (see the results in the next section). In general, however, there exist simply-connected Alexandrov 3-spaces that are not homeomorphic to these connected sums (see [29, Remark 4.1]). A more feasible goal could be to classify the topological type of non-manifold simply-connected Alexandrov spaces with a given number of topological singularities. In this spirit, we recall the following question:

Question 5.2 (Topological Finiteness [34]) Does the class of closed, simply-connected Alexandrov 3-spaces with curv ≥ -1 and diam $\leq D$, for fixed $D > 0$, contain finitely many homeomorphism types?

5.3 Aspherical Spaces and the Borel Conjecture

We have considered simply-connected spaces in the previous section, providing a view of the phenomena that can occur under this assumption in three-dimensional Alexandrov geometry. In a similar way, it is natural to try to understand the topology of other topologically simple classes of spaces such as *aspherical* spaces. Recall that a topological space is *aspherical* if all of its higher homotopy groups vanish. One could, in a certain sense, think of the class of aspherical spaces as complementary to the class of simply-connected spaces.

The theorems we present in this subsection provide a basic picture of the topology of closed and aspherical Alexandrov 3-spaces. One of the main fundamental results in 3-manifold topology is the *Borel conjecture*. Recall that the *Borel conjecture* asserts that if two closed, aspherical n-manifolds are homotopy equivalent, then they must be homeomorphic. The proof of this conjecture for three-dimensional manifolds is a consequence of Perelman's proof of Thurston's geometrization conjecture (see [84, Sect. 2.6]). As we have pointed out in a previous section, a geometrization theorem is available for closed Alexandrov 3-spaces. However, it is unclear whether this implies the Borel conjecture in this more general setting.

In [71, Sect. 6], Núñez-Zimbrón showed that if two closed, aspherical Alexandrov 3-spaces on which the circle acts effectively and isometrically are homotopy equivalent, then they are homeomorphic. The proof of this result is based on the decomposition that any such space X admits as a connected sum of a closed 3-manifold with a finite number of copies of the suspension of $\mathbb{R}P^2$ (see Theorem 6.4). The argument is based on the observation that no connected sum Y of copies of $\mathrm{Susp}(\mathbb{R}P^2)$ can be aspherical, so that a connected sum of the form $M\#Y$, where M is a closed 3-manifold, is aspherical only if $M\#Y \approx M$. Therefore, if X is a closed aspherical Alexandrov 3-space on which the circle acts effectively and isometrically, then the Borel conjecture holds since X must be homeomorphic to a closed 3-manifold. It is immediate to see that the same argument implies, via the connected sum decomposition of Theorem 6.6, that if two closed, aspherical Alexandrov 3-spaces which admit isometric local circle actions (see Sect. 6 below for the definition) are homotopy equivalent, then they are homeomorphic.

A different but related result concerning the validity of the Borel conjecture for Alexandrov 3-spaces was obtained by Bárcenas and the second named author in [6]. In their work they reinforce the condition of asphericity further with a constraint on the Hausdorff measure of the spaces with respect to their diameters, as well as with a topological condition of *irreducibility*, originally defined by Galaz-García, Guijarro and the second named author in [30]. A closed Alexandrov 3-space X is *irreducible* if every embedded 2-sphere in X bounds a 3-ball and, in the case that the set of topologically singular points of X is non-empty, it is further required that every 2-sided $\mathbb{R}P^2$ bounds a cone over a real projective plane $\mathbb{R}P^2$. This definition is fashioned after the definition of irreducibility for 3-manifold, which plays a central role in 3-manifold topology (see, for example, [51]). With this definition in hand we recall the main result of [6].

Theorem 5.3 ([6]) *For any $D > 0$, there exists $\varepsilon = \varepsilon(D) > 0$ such that, if X is a closed, irreducible and aspherical Alexandrov 3-space satisfying*

- $\mathrm{curv}(X) \geq -1$,
- $\mathrm{diam}\, X \leq D$ *and*
- $\mathcal{H}^3(X) \leq \varepsilon$,

then X is homeomorphic to a 3-manifold.

The preceding theorem immediately implies the following corollary which asserts the validity of the Borel conjecture for a certain class of Alexandrov 3-spaces.

Corollary 5.4 ([6]) *For any $D > 0$, there exists $\varepsilon = \varepsilon(D) > 0$ such that, if X_1 and X_2 are closed, irreducible and aspherical Alexandrov 3-spaces satisfying that*

- $\mathrm{curv}(X_i) \geq -1$,
- $\mathrm{diam}\, X_i \leq D$, *and*
- $\mathcal{H}^3(X_i) \leq \varepsilon$,

for $i = 1, 2$, then the Borel conjecture holds for X_1 and X_2, that is, if X_1 is homotopy equivalent to X_2 then X_1 is homeomorphic to X_2.

Corollary 5.4 guarantees that given a homotopy equivalence $f : X_1 \to X_2$, there exists *some* homeomorphism $\tilde{f} : X_1 \to X_2$. In the manifold case, it is true that the initial map f is itself homotopic to a homeomorphism. As the proof of the Borel conjecture in the special case considered in Corollary 5.4 reduces to the Borel conjecture for 3-manifolds, this stronger statement holds true as well.

We now give a rough outline of the proof of Theorem 5.3. By contradiction, let us assume that the thesis does not hold. Then, there exists a sequence of closed, irreducible, and aspherical Alexandrov 3-spaces $\{X_i\}$ with $\mathrm{curv}(X_i) \geq -1$, with uniformly bounded diameters and with their 3-dimensional Hausdorff measures converging to 0. Then Gromov's compactness theorem (see [12, Theorem 10.7.2]) yields that, possibly after passing to a subsequence, there exists a compact Alexandrov space Y (possibly with boundary) such that X_i converges to Y in the Gromov–Hausdorff sense. The Hausdorff dimension of Y must be strictly smaller than 3 as, by the weak convergence of the Hausdorff measures under Gromov–Hausdorff convergence (see [13, Theorem 10.8]), $\mathcal{H}^3(Y) = 0$. At this point we note that, in the terminology of Sect. 7, the sequence $\{X_i\}$ is a *collapsing* sequence (to Y). Mitsuishi and Yamaguchi have classified the topologies of such X_i for large enough i. To proceed with the proof of Theorem 5.3, we use this classification and a case by case analysis of asphericity to obtain a contradiction.

In general it is not known whether any two closed, aspherical Alexandrov 3-spaces that are homotopy equivalent must also be homeomorphic. However, the evidence gathered by the results in this section seems to point towards the following conjecture, which if true, would answer this question in the affirmative.

Conjecture 5.1 Every closed and aspherical Alexandrov 3-space is homeomorphic to a 3-manifold.

We conclude this section by pointing out that the geometry and topology of non-compact, topologically singular, Alexandrov 3-spaces remains to be explored.

6 Alexandrov 3-Spaces with Compact Lie Group Actions

Spaces with large groups of isomorphisms are of interest in different areas of mathematics. In the context of differential geometry, the study of smooth manifolds with smooth actions of compact Lie groups is a subject with a long history (see, for example, [56, 58]) that has brought about further developments in Riemannian geometry. It is therefore natural to consider Alexandrov spaces with isometric group actions and to generalize the theory of compact transformation groups on manifolds [9] to the case of Alexandrov spaces. As in the smooth case, a reasonable starting point in the study of closed Alexandrov spaces with isometric compact Lie group actions is to consider those that support "large" actions. Much work has been done in the Riemannian setting, where this point of view has led to topological and equivariant classification results for smooth manifolds with Riemannian metrics of positive or non-negative sectional curvature, in the context of the *Grove program* (see [41, 43, 98]). One can therefore strive for corresponding results in the context of Alexandrov geometry. There has already been some work in this direction (see, for example, [28, 32, 33, 50, 71]) and we will focus our attention here on results on Alexandrov 3-spaces. For a more general discussion on group actions on Alexandrov spaces the reader may consult [89]. As in the preceding section, we will concentrate on the case where the Alexandrov 3-space is not a manifold.

6.1 Setup

Let X be an Alexandrov space. A bijection $f\colon X \to X$ is an *isometry* if $d(f(p), f(q)) = d(p, q)$ for any pair of points $p, q \in X$. We denote the group of isometries of X by $\mathrm{Isom}(X)$. Fukaya and Yamaguchi showed that $\mathrm{Isom}(X)$ is a Lie group [26]. The corresponding result for Riemannian manifolds was proved by Myers and Steenrod [69]. By a theorem of van Dantzig and van der Waerden [18], if X is compact, then $\mathrm{Isom}(X)$ is also compact. Myers and Steenrod also obtained a sharp upper bound on the dimension of the isometry group of a Riemannian manifold, namely, if $\dim(M) = n$, then $\dim(\mathrm{Isom}(M)) \leq n(n+1)/2$, and obtained a rigidity statement in the equality case (see [69]; cf. [58]), showing that such Riemannian manifolds must be isometric to a sphere, a real projective space, Euclidean space or hyperbolic space. In the context of Alexandrov spaces, Galaz-García and Guijarro obtained the same upper bound on $\dim(\mathrm{Isom}(X))$ and generalized the rigidity result, obtaining the same list of spaces as in the Riemannian case (see [28]).

We will consider actions $G \times X \to X$ of a compact Lie group G on X such that the restriction of the action to sets of the form $\{g\} \times X$ are isometries of X. In this case, one says that the action is *isometric* or that G acts *isometrically* (or *by isometries*) on X.

We will denote the orbit of a point $x \in X$ by $G(x)$, that is,

$$G(x) = \{gx \mid g \in G\}.$$

It is easy to show that $G(x)$ is homeomorphic to G/G_x, where

$$G_x = \{g \in G \mid gx = x\}$$

is the *isotropy subgroup* of x in G. The closed subgroup of G given by $\bigcap_{x\in X} G_x$ is called the *ineffective kernel* of the action. If the ineffective kernel is trivial, we will say that the action is *effective*.

The homeomorphism $G(x) \approx G/G_x$ for each x shows that there is a correspondence between orbits and isotropy groups in the following sense. Given an isotropy subgroup $H \leq G$, one says that $G(x)$ *is of type* (H) if G_x is conjugate to H. **The set of orbit types naturally carries a partial ordering defined as follows.** We say that $(H) \leq (K)$ if K is conjugate to a subgroup of H. One of the main tools in the theory of compact transformation groups is the principal orbit theorem, obtained for Alexandrov spaces by Galaz-García and Guijarro in [28], (see [41] for the Riemannian case).

Theorem 6.1 (Principal Orbit Theorem [28]) *Let G be a compact Lie group acting isometrically on an n-dimensional Alexandrov space X. Then there is a unique maximal orbit type and the orbits with maximal orbit type, the* principal orbits *of the action, form an open and dense subset of X.*

Given a subset $A \subset X$ we denote its image under the orbit projection map $\pi \colon X \to X/G$ by A^*. In particular, $X^* = X/G$. It was proved in [13] (cf. [12, Proposition 10.2.4]) that the orbit space X^* equipped with the distance between orbits is an Alexandrov space with the same lower curvature bound as X. This is a consequence of the fact that the projection $\pi : X \to X^*$ is a *submetry*, that is, π sends balls of radius $r > 0$ in X to balls of radius $r > 0$ in X^*.

Let $x \in X$. Given $A \subset \Sigma_x X$, we define the *set of normal directions to* A as

$$A^{\perp} = \{v \in \Sigma_x X \mid \angle(v, w) = \operatorname{diam}(\Sigma_x X)/2 \text{ for all } w \in A\}.$$

Let S_x denote the tangent unit space to the orbit G/G_x. If $\dim(G/G_x) > 0$ then the set $S_x^{\perp}$ is a compact, totally geodesic Alexandrov subspace of $\Sigma_x X$ with curvature bounded below by 1. Moreover, $\Sigma_x X$ is isometric to the join $S_x * S_x^{\perp}$ with the standard join metric and either $S_x^{\perp}$ is connected or it contains exactly two points at distance π (see [33]).

The slice theorem is a fundamental result in the theory of compact transformation groups and provides a canonical decomposition of a small invariant tubular neighborhood of each orbit as well as more information on how the orbit types are distributed on the space. Several (a posteriori equivalent) definitions of the notion of slice are available depending on the generality. Let us recall the definition of a *slice* from Bredon [9].

A *slice* at a point $x \in X$ is a subset $S \subset X$ containing x which is closed in $G(S)$ and that satisfies the following properties:

- $G(S)$ is an open neighborhood of $G(x)$,
- $G_x(S) = S$, and
- if $(gS) \cap S \neq \varnothing$, then $g \in G_x$.

The existence of a slice at each point can be shown in high generality. Indeed, Montgomery and Yang [67] showed that if a compact Lie group G acts on a completely regular topological space, then there exists a slice at each point. We now state Harvey and Searle's slice theorem for Alexandrov spaces.

Theorem 6.2 (Slice Theorem [50]) *Let G be a compact Lie group acting isometrically on an Alexandrov space X. Then for all $x \in X$, there exists $r_0 > 0$ such that for all $r < r_0$ there is an equivariant homeomorphism*

$$\Phi : G \times_{G_x} K(S_x^{\perp}) \to B_r(G(x)).$$

It is worth noting that, as the slice theorem shows, in the context of Alexandrov spaces the slice at each point can be taken as the cone over $S_x^{\perp}$, where S_x is the unit tangent space to the orbit G/G_x. An important and immediate consequence of the slice theorem is that $\Sigma_{x^*} X^*$, the space of directions at each $x^* \in X^*$, is isometric to $S_x^{\perp}/G_x$.

One can measure the size of the isometry group $\operatorname{Isom}(X)$ of a closed Alexandrov space X by means of different invariants. Three natural ones are the *symmetry degree*, given by $\operatorname{symdeg}(X) = \dim(\operatorname{Isom}(X))$, the *symmetry rank*, given by $\operatorname{symrk}(X) = \operatorname{rank}(\operatorname{Isom}(X))$, and the *cohomogeneity* of the action, defined as the dimension of the orbit space $X/\operatorname{Isom}(X)$. Here we will discuss isometric actions of compact connected Lie groups on closed Alexandrov 3-spaces from the point of view of the cohomogeneity of the action.

Let X be a closed Alexandrov 3-space with an isometric action of a compact connected Lie group G. Thus, the cohomogeneity of the action is 0, 1, 2 or 3. We need not consider the case where the cohomogeneity is three, since this implies that G is the identity.

6.2 Homogeneous Spaces

When the cohomogeneity is 0, X must be a homogeneous space and it follows from work of Berestovskiĭ that X is isometric to a Riemannian manifold (see [7]).

6.3 Cohomogeneity One Spaces

Topological manifolds with cohomogeneity one actions were first studied by Mostert [68] (see also [35]) and classified in dimension three by Mostert [68] and Neumann [70] (see also [35, 53, 76] for the classification in dimensions at most 7 in the topological and smooth categories). The structure of general closed cohomogeneity one Alexandrov spaces is given by the following result, which generalizes the structure result for closed cohomogeneity-one smooth manifolds:

Theorem 6.3 (Cohomogeneity One Alexandrov Spaces [33]) *Let X be a closed Alexandrov space with an effective cohomogeneity one isometric action of a compact connected Lie group G with principal isotropy H. Then the following hold:*

- *The orbit space X/G is homeomorphic to a circle or to a closed interval.*
- *If X/G is a circle, then M is equivariantly homeomorphic to a fiber bundle over $\mathbb{S}^1$ with fiber G/H and structure group $N(H)/H$. In particular, X is a manifold.*
- *If $X/G \approx [-1, +1]$, then there is a group diagram (G, H, K^-, K^+) with*

$$\begin{array}{ccccc} & & G & & \\ & {\scriptstyle j_-}\nearrow & & \nwarrow{\scriptstyle j_+} & \\ K^- & & & & K^+ \\ & {\scriptstyle i_-}\nwarrow & & \nearrow{\scriptstyle i_+} & \\ & & H & & \end{array}$$

  where $K^{\pm}$ are the isotropy groups at ± 1 and $K^{\pm}/H$ are isometric to homogeneous spaces with $\sec > 0$.
- *The space X is the union of two fiber bundles with fiber $C(K^{\pm}/H)$ and base the singular orbits $G/K^{\pm}$.*

- *Conversely, any diagram (G, H, K^-, K^+), such that $K^{\pm}/H$ is a homogeneous space of positive curvature determines an Alexandrov G-space of cohomogeneity* 1.

In Theorem 6.3, if X is a smooth manifold and the action is smooth, then the fibers of the double cone bundle decomposition are disks, i.e., cones over spheres. If X is only assumed to be a topological manifold, then one must consider as fibers, in addition to disks, cones over the Poincaré homology sphere (see [35]).

Let X be a closed cohomogeneity one Alexandrov 3-space. If X is a 3-manifold, then it follows from the work of Mostert and Neumann [68, 70] that X must be one of T^3, $\mathbb{S}^3$, $\mathbb{L}_{p,q}$, $\mathbb{S}^2 \times \mathbb{S}^1$, $\mathbb{S}^2 \tilde{\times} \mathbb{S}^1$, $\mathbb{K}\mathrm{l} \times \mathbb{S}^1$, $\mathbb{R}P^2 \times \mathbb{S}^1$ or $\mathbb{A}$. Here, $\mathbb{S}^2 \tilde{\times} \mathbb{S}^1$ is the non-trivial $\mathbb{S}^2$ bundle over $\mathbb{S}^1$, $\mathbb{L}_{p,q}$ denotes a lens space, and $\mathbb{K}\mathrm{l}$ is the Klein bottle; the space $\mathbb{A}$ is the manifold $\mathbb{M}\mathrm{b} \times \mathbb{S}^1 \cup \mathbb{S}^1 \times \mathbb{M}\mathrm{b}$, where $\mathbb{M}\mathrm{b}$ is the compact Möbius band and the halves $\mathbb{M}\mathrm{b} \times \mathbb{S}^1$, $\mathbb{S}^1 \times \mathbb{M}\mathrm{b}$ intersect canonically in $\mathbb{S}^1 \times \mathbb{S}^1$. If X is not a manifold, then it was proved in [33] that X must be equivariantly homeomorphic to $\mathrm{Susp}(\mathbb{R}P^2)$ with the suspension of the transitive action of SO(3) on $\mathbb{R}P^2$.

6.4 Cohomogeneity Two Spaces

Let X be a closed Alexandrov 3-space with a cohomogeneity two isometric and effective action of a compact connected Lie group. Since the orbits are one-dimensional, the group acting must be the circle S^1. The topological and equivariant classification in the case where X is a manifold follows from the work of Orlik and Raymond [73, 86], who classified the effective actions of the circle on any closed, connected topological 3-manifold M (see [72]). The orbit space of such an action is a topological 2-manifold, possibly with boundary and each equivariant homeomorphism type is determined by a set of invariants

$$(b; (\varepsilon, g, f, t), \{(\alpha_1, \beta_1), \ldots, (\alpha_n, \beta_n)\}).$$

Here, b is the obstruction class (in the sense of obstruction theory, see for example [19, Chap. 7]) for the principal stratum of the action to be a trivial principal S^1-bundle. The symbol ε takes two possible values, corresponding to the orientability of the orbit space. The genus of the orbit space is denoted by g. The number of connected components of the fixed point set is denoted by f, while t is the number of $\mathbb{Z}_2$-isotropy connected components. The pairs $\{(\alpha_i, \beta_i)\}_{i=1}^n$ are the Seifert invariants (see [72] for the definition) associated to the exceptional orbits of the action, if any.

Núñez-Zimbrón carried out the topological and equivariant classification of topologically singular closed Alexandrov 3-spaces with an isometric circle action [71]. Recall that a closed Alexandrov 3-space X that is not a 3-manifold has finitely many topologically singular points, i.e., points whose space of directions is homeomorphic to the real projective plane $\mathbb{R}P^2$. To account for these points, one adds an unordered s-tuple $(r_1, r_2, \ldots, r_s)$ of even positive integers to the set of invariants in the manifold case. The integer s corresponds to the number of boundary components in the orbit space that contain orbits of topologically singular points. The integers r_i correspond to the number of topologically singular points in the i-th boundary component of the orbit space with orbits of topological singularities. If there are no topologically singular points one considers this s-tuple to be empty. The classification is then given by the following theorem. Recall that a homeomorphism $f: X \to Y$ between G-spaces X, Y is *weakly equivariant* if there exists an isomorphism $\varphi: G \to G$ such that $f(gx) = \varphi(g)f(x)$ for all $g \in G$ and all $x \in X$.

Theorem 6.4 (Spaces with Circle Actions [71]) *Let S^1 act effectively and isometrically on a closed Alexandrov 3-space X. Assume that X has $2r$ topologically singular points, $r \geq 0$. Then the following hold:*

(1) *The set of inequivalent (up to weakly equivariant homeomorphism) effective, isometric circle actions on X is in one-to-one correspondence with the set of unordered tuples*

$$\big(b; (\varepsilon, g, f, t); \{(\alpha_i, \beta_i)\}_{i=1}^n; (r_1, r_2, \ldots, r_s)\big)$$

where the permissible values for b, ε, g, f, t and $\{(\alpha_i, \beta_i)\}_{i=1}^n$, are the same as in the manifold case and $(r_1, r_2, \ldots, r_s)$ is an unordered s-tuple of even positive integers r_i such that $r_1 + \ldots + r_s = 2r$.

(2) *X is weakly equivariantly homeomorphic to*

$$M\#\underbrace{\mathrm{Susp}(\mathbb{R}P^2)\#\cdots\#\,\mathrm{Susp}(\mathbb{R}P^2)}_{r}$$

where M is the closed 3-manifold given by the set of invariants

$$\big(b; (\varepsilon, g, f+s, t); \{(\alpha_i, \beta_i)\}_{i=1}^n\big)$$

in the manifold case.

We now outline the main points of the proof of the previous theorem. Observe first that there are different orbit types, which correspond to the possible isotropy groups of the action. These in turn, correspond to the closed subgroups of $\mathbb{S}^1$: the trivial subgroup $\{e\}$, the cyclic subgroups $\mathbb{Z}_k$, $k \geq 2$, and S^1 itself. In particular, since each orbit is homeomorphic to the quotient of S^1 by the corresponding isotropy group, orbits in X are either 0-dimensional or 1-dimensional. This observation and the finiteness of the set of topologically singular points of X imply that topologically singular points are fixed by the action.

We let F be the set of fixed points of the action and let $RF = F \setminus S_X$, the set of topologically regular fixed points. The points whose isotropy is not S^1 are topologically regular, therefore the notion of local orientation makes sense (see for example the remark on orientability in [80, p. 124]). We will say that an orbit with isotropy $\mathbb{Z}_k$ acting without reversing the local orientation is *exceptional*; we will denote the set of points on exceptional orbits by E. An orbit with isotropy $\mathbb{Z}_2$ that acts reversing the local orientation will be called *special exceptional* and the set of points on such orbits will be denoted by SE. The orbits with trivial isotropy will be called *principal*.

An analysis of the structure of X around each orbit via the slice Theorem 6.2 yields the structure of the orbit space X^*: It is a compact 2-manifold possibly with boundary in which the interior points correspond to principal orbits except for a finite number (possibly zero) of points which are associated to exceptional orbits. For each of the boundary components one of the following possibilities occurs: The component consists entirely of RF-orbits, SE-orbits or the component can be decomposed as a union of closed non-trivial intervals with SE or RF isotropy in their interiors and with the endpoints corresponding to orbits of topologically singular points. Note that this implies in particular that X must have an even number of topologically singular points, recovering the result mentioned before Example 3.1 above.

At this point one must show that each possible orbit space corresponds exactly to a single closed Alexandrov 3-space with an isometric circle action up to equivariant homeomorphism. The main tool is a cross-sectioning theorem asserting the existence of a cross-section to the action in the absence of exceptional orbits, which we recall below. This theorem extends the corresponding result of Orlik and Raymond for circle actions on 3-manifolds. Note, however, that in the manifold case one requires that $F \neq \varnothing$ while for non-manifold Alexandrov 3-spaces this is automatically true as topologically singular points are fixed points.

Proposition 6.5 *If S^1 acts effectively and isometrically on a closed, Alexandrov 3-space X with $E = \varnothing$ and $F \neq \varnothing$, then there exists a cross-section to the action.*

This cross-section given by the preceding proposition can be used to build equivariant homeomorphisms between spaces with isomorphic orbit spaces if no exceptional orbits are present. If $E \neq \varnothing$ one can show that the action is completely determined by the restriction of the cross-section to a tubular neighborhood of E coupled with the information given by the Seifert invariants.

To complete the proof of item (1) in Theorem 6.4 we now must show that each admissible orbit space X^* is indeed the orbit space of a closed Alexandrov 3-space with some isometric circle action. One can easily construct a topological space X with a circle action whose orbit space is X^* by gluing together the "building blocks" obtained via the slice theorem. In other words, each small neighborhood of each orbit type in X^* can be "lifted" uniquely (up to equivariant homeomorphism). A more delicate point is to show that this space X indeed admits an Alexandrov metric. This is achieved by using the branched double cover construction for X, obtaining a topological 3-manifold $\tilde{X}$ which doubly covers X up to a finite number of isolated points. The S^1 action on $\tilde{X}$ is equivalent to a smooth action by the work of Orlik and Raymond on circle actions on 3-manifolds. On $\tilde{X}$, then, one can do an averaging procedure as in [5, Theorem 3.65] to obtain an invariant Riemannian metric which has bounded sectional curvature by compactness and which, in turn, can be projected down to X to obtain an orbifold Riemannian metric on X with sectional curvature from bounded below. We refer the reader to [32] for the details.

The connected sum decomposition of X in item (2) of Theorem 6.4 is obtained by considering orbit spaces which are homeomorphic to 2-disks in which $E = \varnothing$ and with at least two topologically singular points. Proposition 6.5 and the slice Theorem 6.2 are then used to prove that these orbit spaces correspond to equivariant connected sums of suspensions of $\mathbb{R}P^2$ with a standard circle action.

6.5 Spaces with Local Circle Actions

A wider class of Alexandrov 3-spaces with symmetry results from generalizing the notion of isometric circle action to that of an isometric *local* circle action. An isometric local circle action on a closed Alexandrov 3-space X is a decomposition of X into disjoint, simple, closed curves, which we call *fibers*, each having a tubular neighborhood which admits an effective, isometric circle action (with respect to the restricted metric of X) whose orbits are the curves of the decomposition. We do not exclude the possibility that some of the curves in the decomposition consist of single points. In this case the equivariant and topological classification was obtained by the authors in [32], generalizing the corresponding classifications for closed 3-manifolds obtained by Fintushel [25] and Orlik and Raymond [74].

Several new invariants must be added to those in Theorem 6.4 to account for the fact that tubular neighborhoods of each type of (one-dimensional) orbit may not be orientable, or equivalently, the boundary of such a tubular neighborhood may not be homeomorphic to a 2-torus but, rather, to a Klein bottle. Let us present the classification theorem, followed by the explanation of the invariants appearing in it.

Theorem 6.6 (Spaces with Local Circle Actions [32]) *Let X be a closed Alexandrov 3-space with a local isometric S^1-action. If X has $2r \geq 0$ topologically singular points, then the following hold:*

(1) *Isometric local circle actions (up to equivariant equivalence) are in one-to-one correspondence with unordered tuples*

$$\{b; \varepsilon, g, (f, k_1), (t, k_2), (s, k_3); \{(\alpha_i, \beta_i)\}_{i=1}^n; (r_1, r_2, \ldots, r_{s-k_3}); (q_1, q_2, \ldots, q_{k_3})\},$$

where the admissible values for b, ε, g, (f, k_1), (t, k_2) and (α_i, β_i) are the same as in the manifold case, and $(r_1, r_2, \ldots, r_{s-k_3})$ and $(q_1, q_2, \ldots, q_{k_3})$ are unordered $(s - k_3)$- and k_3-tuples of even non-negative integers r_i, q_j, respectively, such that $r_1 + \ldots + r_{s-k_3} + q_1 + \ldots q_{k_3} = 2r$.

(2) *There is an equivariant equivalence of X with*

$$M\#\underbrace{\mathrm{Susp}(\mathbb{R}P^2)\#\cdots\#\mathrm{Susp}(\mathbb{R}P^2)}_{r \text{ summands}},$$

where M is the closed 3-manifold determined by the set of invariants

$$\{b; \varepsilon, g, (f+s, k_1+k_3), (t, k_2); \{(\alpha_i, \beta_i)\}_{i=1}^n\}$$

in the manifold case.

Let us now explain the invariants appearing in Theorem 6.6. As in the case of global circle actions, there are several types of fibers, each corresponding to the possible isotropy groups. The fiber types F, E and SE are defined as in the case of global circle actions in the preceding subsection. We denote principal fibers by R and orbits of topologically singular points by SF. Similarly as well, the fiber space X^* is a topological 2-manifold possibly with boundary. When present, the boundary is composed of the images of F-, SE- and topologically singular fibers while the interior of X^* consists of R-fibers and a finite number of E-fibers.

A closed Alexandrov 3-space with an isometric local circle action can be decomposed into the following pieces:

(a) *building blocks* which arise by considering small tubular neighborhoods of connected components of fibers of type F, SF, E and SE;
(b) an S^1-fiber bundle (composed only of R-fibers) with structure group $\mathrm{O}(2)$ over a compact 2-manifold with boundary, which corresponds to the complement in X of the union of the building blocks in part (a).

While a similar statement is true in the case of global circle actions, the main differences in the case of local circle actions are, on the one hand, the possible building blocks that appear when examining the neighborhoods of each fiber via the slice theorem and, on the other hand, that in the global case the structure group of the fiber bundle in item (b) can be reduced to $\mathrm{SO}(2)$.

A building block is called *simple* if its boundary is homeomorphic to a torus, and *twisted* if its boundary is homeomorphic to a Klein bottle (see [32, Sect. 3]). A pair (ε, k) where k is a non-negative, even integer and ε takes one of six possible symbolic values can be uniquely associated to the $\mathrm{O}(2)$-bundle of R-fibers, completely characterizing it up to *weak equivalence of bundles* (see [25, Sect. 1]). We denote the genus of X^* by $g \geq 0$. We let f, t, k_1, k_2 be non-negative integers such that $k_1 \leq f$ and $k_2 \leq t$, where k_1 is the number of twisted F-blocks and k_2 is the number of twisted SE-blocks. The number $f - k_1$ is the number of simple F-blocks and $t - k_2$ is the number of simple SE-blocks. A non-negative integer n will denote the number of E-fibers and we let $\{(\alpha_i, \beta_i)\}_{i=1}^n$ be the corresponding Seifert invariants. The invariant b is a certain obstruction class defined in similar fashion to that of the global circle actions case (see [74, Sect. 2] for a precise definition). We let s, k_3 be non-negative integers, where $k_3 \leq s$ is the number of twisted SF-blocks. Hence $s - k_3$ is the number of simple SF-blocks, and we let $(r_1, r_2, \ldots, r_{s-k_3})$ and $(q_1, q_2, \ldots, q_{k_3})$ be $(s-k_3)$- and k_3-tuples of non-negative even integers corresponding to the number of topologically singular points in each simple and twisted SF-block, respectively. The numbers k, k_1, k_2, and k_3 satisfy $k_1 + k_2 + k_3 = k$.

It is worth remarking here that all Seifert manifolds admit a local circle action, but not necessarily a global circle action. Prior to the work of Orlik and Raymond [73, 74], a classification of Seifert manifolds in terms of symbolic and numeric invariants was obtained by Seifert in [90]. An analogous classification for generalized Seifert fibered spaces (see Sect. 7 below) as defined by Mitsuishi and Yamaguchi [66] is still unknown, as is the precise relation such a classification would share with that of the local circle actions on closed Alexandrov 3-spaces of Theorem 6.6.

7 Collapse

We have already mentioned a few instances of collapse in three-dimensional Alexandrov geometry. Let us recall here the definition of collapse. Let $\mathcal{A}^n_{k,D}$ be the class of n-dimensional Alexandrov spaces with lower curvature bound k and diameter bounded above by D. By Gromov's compactness theorem (see [12, Theorem 10.7.2]), $\mathcal{A}^n_{k,D}$ is compact with respect to the topology induced by the Gromov–Hausdorff distance. Let $\{X^n_i\} \subset \mathcal{A}^n_{k,D}$ be a Gromov–Hausdorff converging sequence with limit X. If X is n-dimensional, then, by Perelman's stability theorem, the spaces X^n_i and X are homeomorphic for sufficiently large i (see [57, 79]). If the limit X is lower-dimensional, then the sequence $\{X^n_i\}$ *collapses*. As noted in Sect. 2, the collapse of Riemannian manifolds in $\mathcal{M}^n_{k,D} \subset \mathcal{A}^n_{k,D}$ is still not well-understood and this question can be explored in the context of the larger class $\mathcal{A}^n_{k,D}$. Observe that closed Alexandrov 3-spaces with an effective, isometric circle action fall within the class $\mathcal{A}^3_{k,D}$.

Shioya and Yamaguchi [92, 93] examined the collapse of Riemannian manifolds in $\mathcal{M}^3_{k,D}$ and determined the possible topology of the elements M_i of the sequence $\{M_i\} \subset \mathcal{M}^3_{k,D}$, for large i, according to the topology of the limit X. More recently, Mitsuishi and Yamaguchi [66] considered the collapse of Alexandrov spaces in $\mathcal{A}^3_{k,D}$. In the case of collapsing sequences $\{X_i\} \subset \mathcal{A}^3_{k,D}$, for sufficiently large i, the collapsing spaces X_i can be written in terms of basic building blocks: so-called *generalized Seifert fiber spaces* (which are similar to Seifert spaces but with some possibly singular fibers), *generalized solid tori*, *generalized Klein bottles* and other, more familiar, spaces, such as interval bundles over the Klein bottle. The structure of collapsed Alexandrov 3-spaces is obtained by a careful analysis of the limit X, which can be zero-, one- or two-dimensional. It follows that X is a point, a closed interval, a circle or an Alexandrov surface with or without boundary. As in the case of Alexandrov 3-spaces with group actions, where the structure of the 3-space can be recovered from that of its orbit space, the topological and metric structure of X determines the structure of X_i for sufficiently large i. For example, when X is one-dimensional, i.e., a closed interval, for large i the spaces X_i are the union of two pieces whose topology can be explicitly determined (see [66, Theorem 1.8] and compare with Theorem 6.3). The general situation for both collapsed Riemannian

3-manifolds and collapsed Alexandrov 3-spaces is rather intricate, though, so we refer the reader to the original articles [66, 92, 93] for precise statements and proofs of these important structure theorems.

Using Mitsuishi and Yamaguchi's classification of collapsing Alexandrov 3-spaces of as well as the classification of local circle actions, Guijarro and the authors obtained a geometrization result for sufficiently collapsed Alexandrov 3-spaces [30]. Roughly speaking, they showed that a closed, irreducible and sufficiently collapsed Alexandrov 3-space X is modeled in one of the eight Thurston geometries (excluding the hyperbolic geometry $\mathbb{H}^3$).

Theorem 7.1 (Geometrization of Sufficiently Collapsed Alexandrov 3-Spaces [30]) *For any $D > 0$ there exists $\varepsilon = \epsilon(D) > 0$ such that if X is a closed, irreducible Alexandrov 3-space*

- $\mathrm{curv}(X) \geq -1$,
- $\mathrm{diam}(X) \leq D$ *and*
- $\mathcal{H}^3(X) \leq \varepsilon$,

then X admits a geometric structure modeled on $\mathbb{S}^3$, $\mathbb{S}^2 \times \mathbb{R}$, $\mathbb{H}^2 \times \mathbb{R}$, $\widetilde{\mathrm{SL}_2(\mathbb{R})}$, Nil *or* Sol.

This result extends part of the work of Shioya and Yamaguchi in [92], formulated in the manifold case, to Alexandrov spaces. The exclusion of hyperbolic geometry in Theorem 7.1 is granted by a result of independent interest proved in [30, Remark B], namely, that a closed collapsing Alexandrov 3-space cannot admit hyperbolic geometry. We refer the reader to [30] and the recent survey [31] for additional details.

Acknowledgments These notes are based on a minicourse given by the first named author at the 11th Minimeeting on Differential Geometry held at CIMAT, in Guanajuato, Mexico, on December 10–12, 2018. These talks were in turn based on [27] and the present notes are an expanded and updated version of this earlier survey. Both authors would like to thank Rafael Herrera, Luis Hernández-Lamoneda, and Gerardo Arizmendi, who organized the meeting, as well as CIMAT, for their hospitality and their support in preparing these notes. The authors would also like to thank Luis Guijarro, with whom most of the joint work presented in this survey was carried out. The first named author would like to thank John Lott for bringing to his attention reference [55] on singular 3-manifolds.

References

1. S. Alexander, V. Kapovitch, and A. Petrunin, *Alexandrov geometry*, Book draft, http://anton-petrunin.github.io/book/arXiv.pdf, February 1, 2021.
2. S. B. Alexander and R. L. Bishop, *Curvature bounds for warped products of metric spaces*, Geom. Funct. Anal. **14** (2004), no. 6, 1143–1181. MR 2135163
3. Stephanie Alexander, Vitali Kapovitch, and Anton Petrunin, *An invitation to Alexandrov geometry*, SpringerBriefs in Mathematics, Springer, Cham, 2019, CAT(0) spaces. MR 3930625

4. Stephanie B. Alexander and Richard L. Bishop, *Warped products admitting a curvature bound*, Adv. Math. **303** (2016), 88–122. MR 3552521
5. Marcos M. Alexandrino and Renato G. Bettiol, *Lie groups and geometric aspects of isometric actions*, Springer, Cham, 2015. MR 3362465
6. Noé Bárcenas and Jesús Núñez-Zimbrón, *On topological rigidity of alexandrov 3-spaces*, Revista Matemática Iberoamericana (2020).
7. V. N. Berestovskiĭ, *Homogeneous manifolds with an intrinsic metric. II*, Sibirsk. Mat. Zh. **30** (1989), no. 2, 14–28, 225. MR 997464
8. J. Bertrand, C Ketterer, Ilaria Mondello, and T. Richard, *Stratified spaces and synthetic Ricci curvature bounds*, arXiv:1804.08870 [math.DG] (2018).
9. Glen E. Bredon, *Introduction to compact transformation groups*, Academic Press, New York-London, 1972, Pure and Applied Mathematics, Vol. 46. MR 0413144
10. Martin R. Bridson and André Haefliger, *Metric spaces of non-positive curvature*, Grundlehren der Mathematischen Wissenschaften [Fundamental Principles of Mathematical Sciences], vol. 319, Springer-Verlag, Berlin, 1999. MR 1744486
11. S. V. Bujalo, *Shortest paths on convex hypersurfaces of a Riemannian space*, Zap. Naučn. Sem. Leningrad. Otdel. Mat. Inst. Steklov. (LOMI) **66** (1976), 114–132, 207, Studies in topology, II. MR 0643664
12. Dmitri Burago, Yuri Burago, and Sergei Ivanov, *A course in metric geometry*, Graduate Studies in Mathematics, vol. 33, American Mathematical Society, Providence, RI, 2001. MR 1835418
13. Yu. Burago, M. Gromov, and G. Perel'man, *A. D. Aleksandrov spaces with curvatures bounded below*, Uspekhi Mat. Nauk **47** (1992), no. 2(284), 3–51, 222. MR 1185284
14. S. V. Buyalo, *Shortest paths on convex hypersurfaces of a Riemannian space*, J. Soviet Math. **12** (1979), 73–85.
15. J. W. Cannon, *Shrinking cell-like decompositions of manifolds. Codimension three*, Ann. of Math. (2) **110** (1979), no. 1, 83–112. MR 541330
16. Mark Cassorla, *Approximating compact inner metric spaces by surfaces*, Indiana Univ. Math. J. **41** (1992), no. 2, 505–513. MR 1183356
17. Fabio Cavalletti and Emanuel Milman, *The globalization theorem for the curvature dimension condition*, Invent. Math., in press.
18. D. van Dantzig and B. L. van der Waerden, *Über metrisch homogene räume*, Abh. Math. Sem. Univ. Hamburg **6** (1928), no. 1, 367–376. MR 3069509
19. James F. Davis and Paul Kirk, *Lecture notes in algebraic topology*, Graduate Studies in Mathematics, vol. 35, American Mathematical Society, Providence, RI, 2001. MR 1841974
20. Qintao Deng, Fernando Galaz-García, Luis Guijarro, and Michael Munn, *Three-dimensional Alexandrov spaces with positive or nonnegative Ricci curvature*, Potential Anal. **48** (2018), no. 2, 223–238. MR 3748392
21. Jonathan Dinkelbach and Bernhard Leeb, *Equivariant Ricci flow with surgery and applications to finite group actions on geometric 3-manifolds*, Geom. Topol. **13** (2009), no. 2, 1129–1173. MR 2491658
22. Manfredo Perdigão do Carmo, *Riemannian geometry*, Mathematics: Theory & Applications, Birkhäuser Boston, Inc., Boston, MA, 1992, Translated from the second Portuguese edition by Francis Flaherty. MR 1138207
23. Robert D. Edwards, *The topology of manifolds and cell-like maps*, Proceedings of the International Congress of Mathematicians (Helsinki, 1978), Acad. Sci. Fennica, Helsinki, 1980, pp. 111–127. MR 562601
24. ——, *Suspensions of homology spheres*, arXiv:math/0610573 [math.GT] (2006).
25. Ronald Fintushel, *Local S^1 actions on 3-manifolds*, Pacific J. Math. **66** (1976), no. 1, 111–118. MR 515868
26. Kenji Fukaya and Takao Yamaguchi, *Isometry groups of singular spaces*, Math. Z. **216** (1994), no. 1, 31–44. MR 1273464
27. Fernando Galaz-García, *A glance at three-dimensional Alexandrov spaces*, Front. Math. China **11** (2016), no. 5, 1189–1206. MR 3547925

28. Fernando Galaz-Garcia and Luis Guijarro, *Isometry groups of Alexandrov spaces*, Bull. Lond. Math. Soc. **45** (2013), no. 3, 567–579. MR 3065026
29. ——, *On three-dimensional Alexandrov spaces*, Int. Math. Res. Not. IMRN (2015), no. 14, 5560–5576. MR 3384449
30. Fernando Galaz-García, Luis Guijarro, and Jesús Núñez Zimbrón, *Sufficiently collapsed irreducible Alexandrov 3-spaces are geometric*, Indiana Univ. Math. J. **69** (2020), no. 3, 977–1005. MR 4095180
31. Fernando Galaz-García, Luis Guijarro, and Jesús Núñez-Zimbrón, *Collapsed 3-dimensional alexandrov spaces: A brief survey*, Differential Geometry in the Large, Cambridge University Press, October 2020, pp. 291–310.
32. Fernando Galaz-García and Jesús Núñez Zimbrón, *Three-dimensional Alexandrov spaces with local isometric circle actions*, Kyoto J. Math. **60** (2020), no. 3, 801–823. MR 4134350
33. Fernando Galaz-Garcia and Catherine Searle, *Cohomogeneity one Alexandrov spaces*, Transform. Groups **16** (2011), no. 1, 91–107. MR 2785496
34. Fernando Galaz-García and Wilderich Tuschmann, *Finiteness and realization theorems for Alexandrov spaces with bounded curvature*, Bol. Soc. Mat. Mex. (3) **26** (2020), no. 2, 749–756. MR 4110479
35. Fernando Galaz-García and Masoumeh Zarei, *Cohomogeneity one topological manifolds revisited*, Math. Z. **288** (2018), no. 3–4, 829–853. MR 3778980
36. Nicola Gigli, *Nonsmooth differential geometry—an approach tailored for spaces with Ricci curvature bounded from below*, Mem. Amer. Math. Soc. **251** (2018), no. 1196, v+161. MR 3756920
37. Nicola Gigli and Enrico Pasqualetto, *Lectures on nonsmooth differential geometry*, Springer, Cham, 2020.
38. José Carlos Gómez-Larrañaga, Francisco González-Acuña, and Wolfgang Heil, *Amenable category of three-manifolds*, Algebr. Geom. Topol. **13** (2013), no. 2, 905–925. MR 3044596
39. Misha Gromov, *Dirac and Plateau billiards in domains with corners*, Cent. Eur. J. Math. **12** (2014), no. 8, 1109–1156. MR 3201312
40. ——, *A dozen problems, questions and conjectures about positive scalar curvature*, Foundations of mathematics and physics one century after Hilbert, Springer, Cham, 2018, pp. 135–158. MR 3822551
41. Karsten Grove, *Geometry of, and via, symmetries*, Conformal, Riemannian and Lagrangian geometry (Knoxville, TN, 2000), Univ. Lecture Ser., vol. 27, Amer. Math. Soc., Providence, RI, 2002, pp. 31–53. MR 1922721
42. ——, *Finiteness theorems in Riemannian geometry*, Explorations in complex and Riemannian geometry, Contemp. Math., vol. 332, Amer. Math. Soc., Providence, RI, 2003, pp. 101–120. MR 2018334
43. ——, *Developments around positive sectional curvature*, Surveys in differential geometry. Vol. XIII. Geometry, analysis, and algebraic geometry: forty years of the Journal of Differential Geometry, Surv. Differ. Geom., vol. 13, Int. Press, Somerville, MA, 2009, pp. 117–133. MR 2537084
44. Karsten Grove and Peter Petersen, *Manifolds near the boundary of existence*, J. Differential Geom. **33** (1991), no. 2, 379–394. MR 1094462
45. ——, *A radius sphere theorem*, Invent. Math. **112** (1993), no. 3, 577–583. MR 1218324
46. Karsten Grove and Burkhard Wilking, *A knot characterization and 1-connected nonnegatively curved 4-manifolds with circle symmetry*, Geom. Topol. **18** (2014), no. 5, 3091–3110. MR 3285230
47. Stephanie Halbeisen, *On tangent cones of Alexandrov spaces with curvature bounded below*, Manuscripta Math. **103** (2000), no. 2, 169–182. MR 1796313
48. Richard S. Hamilton, *Three-manifolds with positive Ricci curvature*, J. Differential Geometry **17** (1982), no. 2, 255–306. MR 664497
49. ——, *Four-manifolds with positive curvature operator*, J. Differential Geom. **24** (1986), no. 2, 153–179. MR 862046

50. John Harvey and Catherine Searle, *Orientation and symmetries of Alexandrov spaces with applications in positive curvature*, J. Geom. Anal. **27** (2017), no. 2, 1636–1666. MR 3625167
51. John Hempel, *3-manifolds*, AMS Chelsea Publishing, Providence, RI, 2004, Reprint of the 1976 original. MR 2098385
52. Morris W. Hirsch and Stephen Smale, *On involutions of the* 3*-sphere*, Amer. J. Math. **81** (1959), 893–900. MR 111044
53. Corey A. Hoelscher, *Classification of cohomogeneity one manifolds in low dimensions*, Pacific J. Math. **246** (2010), no. 1, 129–185. MR 2645881
54. Cynthia Hog-Angeloni and Wolfgang Metzler, *Geometric aspects of two-dimensional complexes*, Two-dimensional homotopy and combinatorial group theory, London Math. Soc. Lecture Note Ser., vol. 197, Cambridge Univ. Press, Cambridge, 1993, pp. 1–50. MR 1279175
55. Cynthia Hog-Angeloni and Allan J. Sieradski, *(Singular)* 3*-manifolds*, Two-dimensional homotopy and combinatorial group theory, London Math. Soc. Lecture Note Ser., vol. 197, Cambridge Univ. Press, Cambridge, 1993, pp. 251–280. MR 1279182
56. Wu-Yi Hsiang, *Lie transformation groups and differential geometry*, Differential geometry and differential equations (Shanghai, 1985), Lecture Notes in Math., vol. 1255, Springer, Berlin, 1987, pp. 34–52. MR 895396
57. Vitali Kapovitch, *Perelman's stability theorem*, Surveys in differential geometry. Vol. XI, Surv. Differ. Geom., vol. 11, Int. Press, Somerville, MA, 2007, pp. 103–136. MR 2408265
58. Shoshichi Kobayashi, *Transformation groups in differential geometry*, Classics in Mathematics, Springer-Verlag, Berlin, 1995, Reprint of the 1972 edition. MR 1336823
59. Kyung Whan Kwun and Jeffrey L. Tollefson, PL *involutions of* $S^1 \times S^1 \times S^1$, Trans. Amer. Math. Soc. **203** (1975), 97–106. MR 370634
60. Nan Li, *Globalization with probabilistic convexity*, J. Topol. Anal. **7** (2015), no. 4, 719–735. MR 3400128
61. G. R. Livesay, *Involutions with two fixed points on the three-sphere*, Ann. of Math. (2) **78** (1963), 582–593. MR 155323
62. John Lott and Cédric Villani, *Ricci curvature for metric-measure spaces via optimal transport*, Ann. of Math. (2) **169** (2009), no. 3, 903–991. MR 2480619
63. E. Luft and D. Sjerve, *Involutions with isolated fixed points on orientable* 3*-dimensional flat space forms*, Trans. Amer. Math. Soc. **285** (1984), no. 1, 305–336. MR 748842
64. Sergei Matveev and Dale Rolfsen, *Zeeman's collapsing conjecture*, Two-dimensional homotopy and combinatorial group theory, London Math. Soc. Lecture Note Ser., vol. 197, Cambridge Univ. Press, Cambridge, 1993, pp. 335–364. MR 1279185
65. Ayato Mitsuishi, *Orientability and fundamental classes of alexandrov spaces with applications*, arXiv:1610.08024 (2016).
66. Ayato Mitsuishi and Takao Yamaguchi, *Collapsing three-dimensional closed Alexandrov spaces with a lower curvature bound*, Trans. Amer. Math. Soc. **367** (2015), no. 4, 2339–2410. MR 3301867
67. D. Montgomery and C. T. Yang, *The existence of a slice*, Ann. of Math. (2) **65** (1957), 108–116. MR 87036
68. Paul S. Mostert, *On a compact Lie group acting on a manifold*, Ann. of Math. (2) **65** (1957), 447–455. MR 85460
69. S. B. Myers and N. E. Steenrod, *The group of isometries of a Riemannian manifold*, Ann. of Math. (2) **40** (1939), no. 2, 400–416. MR 1503467
70. Walter D. Neumann, 3*-dimensional* G*-manifolds with* 2*-dimensional orbits*, Proc. Conf. on Transformation Groups (New Orleans, La., 1967), Springer, New York, 1968, pp. 220–222. MR 0245043
71. Jesús Núñez Zimbrón, *Closed three-dimensional Alexandrov spaces with isometric circle actions*, Tohoku Math. J. (2) **70** (2018), no. 2, 267–284. MR 3810241
72. Peter Orlik, *Seifert manifolds*, Lecture Notes in Mathematics, Vol. 291, Springer-Verlag, Berlin-New York, 1972. MR 0426001

73. Peter Orlik and Frank Raymond, *Actions of* SO(2) *on 3-manifolds*, Proc. Conf. on Transformation Groups (New Orleans, La., 1967), Springer, New York, 1968, pp. 297–318. MR 0263112
74. ——, *On* 3*-manifolds with local* SO(2) *action*, Quart. J. Math. Oxford Ser. (2) **20** (1969), 143–160. MR 266214
75. Yukio Otsu and Takashi Shioya, *The Riemannian structure of Alexandrov spaces*, J. Differential Geom. **39** (1994), no. 3, 629–658. MR 1274133
76. Jeff Parker, 4*-dimensional* G*-manifolds with* 3*-dimensional orbits*, Pacific J. Math. **125** (1986), no. 1, 187–204. MR 860758
77. G. Ya. Perel'man, *Elements of Morse theory on Aleksandrov spaces*, Algebra i Analiz **5** (1993), no. 1, 232–241. MR 1220498
78. G. Ya. Perel'man and A. M. Petrunin, *Extremal subsets in Aleksandrov spaces and the generalized Liberman theorem*, Algebra i Analiz **5** (1993), no. 1, 242–256. MR 1220499
79. Grigori Perelman, *Alexandrov spaces with curvatures bounded from below ii*, preprint (1991).
80. A. Petrunin, *Parallel transportation for Alexandrov space with curvature bounded below*, Geom. Funct. Anal. **8** (1998), no. 1, 123–148. MR 1601854
81. Anton Petrunin, *Applications of quasigeodesics and gradient curves*, Comparison geometry (Berkeley, CA, 1993–94), Math. Sci. Res. Inst. Publ., vol. 30, Cambridge Univ. Press, Cambridge, 1997, pp. 203–219. MR 1452875
82. ——, *Semiconcave functions in Alexandrov's geometry*, Surveys in differential geometry. Vol. XI, Surv. Differ. Geom., vol. 11, Int. Press, Somerville, MA, 2007, pp. 137–201. MR 2408266
83. Conrad Plaut, *Metric spaces of curvature* $\geq k$, Handbook of geometric topology, North-Holland, Amsterdam, 2002, pp. 819–898. MR 1886682
84. Joan Porti, *Geometrization of three manifolds and Perelman's proof*, Rev. R. Acad. Cienc. Exactas Fís. Nat. Ser. A Mat. RACSAM **102** (2008), no. 1, 101–125. MR 2416241
85. Frank Quinn, *Presentations and* 2*-complexes, fake surfaces and singular* 3*-manifolds*, Virginia Polytechnic Institute, Blackburg Va. (1981), preprint.
86. Frank Raymond, *Classification of the actions of the circle on* 3*-manifolds*, Trans. Amer. Math. Soc. **131** (1968), 51–78. MR 219086
87. Xiaochun Rong, *Collapsed manifolds with bounded sectional curvature and applications*, Surveys in differential geometry. Vol. XI, Surv. Differ. Geom., vol. 11, Int. Press, Somerville, MA, 2007, pp. 1–23. MR 2408262
88. Peter Scott, *The geometries of* 3*-manifolds*, Bull. London Math. Soc. **15** (1983), no. 5, 401–487. MR 705527
89. Catherine Searle, *An introduction to isometric group actions with applications to spaces with curvature bounded from below*, Geometry of manifolds with non-negative sectional curvature, Lecture Notes in Math., vol. 2110, Springer, Cham, 2014, pp. 21–43. MR 3329928
90. H. Seifert, *Topologie Dreidimensionaler Gefaserter Räume*, Acta Math. **60** (1933), no. 1, 147–238. MR 1555366
91. Katsuhiro Shiohama, *An introduction to the geometry of Alexandrov spaces*, Lecture Notes Series, vol. 8, Seoul National University, Research Institute of Mathematics, Global Analysis Research Center, Seoul, 1993. MR 1320267
92. Takashi Shioya and Takao Yamaguchi, *Collapsing three-manifolds under a lower curvature bound*, J. Differential Geom. **56** (2000), no. 1, 1–66. MR 1863020
93. ——, *Volume collapsed three-manifolds with a lower curvature bound*, Math. Ann. **333** (2005), no. 1, 131–155. MR 2169831
94. Christina Sormani, *Scalar curvature and intrinsic flat convergence*, Measure theory in non-smooth spaces, Partial Differ. Equ. Meas. Theory, De Gruyter Open, Warsaw, 2017, pp. 288–338. MR 3701743
95. Karl-Theodor Sturm, *On the geometry of metric measure spaces. I*, Acta Math. **196** (2006), no. 1, 65–131. MR 2237206
96. Karl-Theodor Sturm, *On the geometry of metric measure spaces. II*, Acta Math. **196** (2006), no. 1, 133–177. MR 2237207

97. Giona Veronelli, *Scalar curvature via local extent*, Anal. Geom. Metr. Spaces **6** (2018), no. 1, 146–164. MR 3884658
98. Burkhard Wilking, *Nonnegatively and positively curved manifolds*, Surveys in differential geometry. Vol. XI, Surv. Differ. Geom., vol. 11, Int. Press, Somerville, MA, 2007, pp. 25–62. MR 2408263
99. Takumi Yokota, *A rigidity theorem in Alexandrov spaces with lower curvature bound*, Math. Ann. **353** (2012), no. 2, 305–331. MR 2915538
100. ——, *On the spread of positively curved Alexandrov spaces*, Math. Z. **277** (2014), no. 1–2, 293–304. MR 3205773
101. Wolfgang Ziller, *Riemannian manifolds with positive sectional curvature*, Geometry of manifolds with non-negative sectional curvature, Lecture Notes in Math., vol. 2110, Springer, Cham, 2014, pp. 1–19. MR 3329927

Topological and Geometric Rigidity for Spaces with Curvature Bounded Below

Jesús Núñez-Zimbrón

1 Introduction

The question of topological or geometric rigidity of spaces is a classical one which has been tackled in a wide array of contexts within mathematics. By topological rigidity we refer to the phenomenon in which two spaces which are homotopy equivalent are automatically homeomorphic. Similarly, geometric rigidity refers to the situation in which two homeomorphic (or diffeomorphic) spaces are automatically isometric. As a motivation, let us recall a few important instances of these phenomena.

It is for example, a central and classical problem in 3-manifold topology to determine whether two given manifolds having the same fundamental group are indeed homeomorphic. A very prominent instance of this problem is the famous Poincaré Conjecture, known to hold true in all dimensions by the work of Smale, Freedman and Perelman, [21, 51–53, 62, 63].

Theorem 1.1 (Poincaré Conjecture) *Every closed n-manifold which is homotopy equivalent to the n-sphere is homeomorphic to it.*

In general an n-manifold is not determined by its fundamental group in the sense that there exist manifolds of the same dimension which are homotopy equivalent (hence have isomorphic fundamental groups) and which are non-homeomorphic (an example of this sort in dimension 3 was first produced in [1]). Nevertheless, there exist families of 3-manifolds which are topologically rigid. Recall that a manifold is aspherical if its higher homotopy groups vanish.

J. Núñez-Zimbrón (✉)
Centro de Ciencias Matemáticas UNAM, Morelia, Mexico
e-mail: zimbron@matmor.unam.mx

G. Arizmendi Echegaray et al. (eds.), *Recent Advances in Alexandrov Geometry*, CIMAT Lectures in Mathematical Sciences,
https://doi.org/10.1007/978-3-030-99298-9_3

Theorem 1.2 (Borel Conjecture) *If two aspherical closed n-manifolds are homotopy equivalent then they are homeomorphic.*

This conjecture is known to hold in dimensions less than or equal to 3 by the classification of compact surfaces and the resolution of Thurston's Geometrization Conjecture by Perelman, [51–53]. In higher dimensions, the Borel Conjecture is implied by the Farrell-Jones conjecture (see [40] and references therein).

An outstanding achievement in Differential Geometry was the discovery of exotic manifolds by Milnor, [42], that is, classes of manifolds which are homeomorphic to one another but which are non-diffeomorphic. A fortiori, this implies that there exist examples of Riemannian n-manifolds which are homotopy equivalent, homeomorphic or diffeomorphic and which are not isometric. However, Mostow showed that geometric rigidity is still valid for certain Riemannian manifolds, such as hyperbolic manifolds, [45].

Theorem 1.3 (Mostow Rigidity) *Let M and N be closed hyperbolic n-manifolds with $n \geq 3$. If M is homotopy equivalent to N then M and N are isometric.*

In fact this geometric rigidity was greatly generalized in the work of Besson, Courtois, Gallot in [8, 9], relaxing the curvature assumption by using the so-called volume entropy h (see Sect. 4 for the definition).

Theorem 1.4 *Let (N, g_{hyp}) be a closed hyperbolic n-manifold, (M, g) a Riemannian n-manifold and $f : M \to N$ a continuous map of non-zero degree. Then*

(i) $h^n(g)\,\mathrm{vol}(M, g) \geq |\deg f| h^n(g_{hyp})\,\mathrm{vol}(N, g_{hyp})$
(ii) The equality occurs if and only if f is homotopic to a $(\deg f)$*-degree Riemannian cover.*

Volume entropy is by itself worth studying as it is related to several important quantities such as minimal volume and simplicial volume. In fact there is a geometric rigidity under the assumption of maximal volume entropy for manifolds with Ricci curvature bounded below due to Ledrappier and Wang [37].

Theorem 1.5 *Let (M^n, g) be a compact Riemannian manifold with $Ric \geq -(n-1)$. Then the volume entropy of M satisfies $h(M, g) \leq (n-1)$ and equality holds if and only if M is hyperbolic.*

In some sense the natural objects of study in geometry are metric spaces, not only Riemannian manifolds. Nevertheless one usually assumes some regularizing conditions as a generic metric space can be extremely wild. Curvature bounds usually impose severe restrictions on the topology and geometry of spaces and therefore act as regularizing conditions. In the cases of sectional and Ricci curvature lower bounds they can be defined in a synthetic context without the need for a differential structure.

Alexandrov spaces are geodesic metric spaces which admit a lower sectional curvature bound in a synthetic sense via comparison of distances in geodesic triangles. In a similar way, $\mathsf{RCD}(K, N)$ and $\mathsf{RCD}^*(K, N)$ spaces can be thought of as the synthetic analog to Ricci curvature being bounded below by K, for dimension at most N, defined by the convexity of the entropy functional of measures over the space (see Sect. 2 for the precise definitions of these classes of spaces).

It is then natural to wonder whether similar topological and geometric results occur in these classes of spaces. Some rigidity results are already available for these spaces: the Poincaré Conjecture as well as Thurston's Geometrization were generalized for Alexandrov 3-spaces by Galaz-García and Guijarro in [23].

In this article we briefly survey two recent developments in the understanding of the topological and geometric rigidity of singular spaces with curvature bounded below: We go over the recent results in [46] and [7] regarding the Borel Conjecture for Alexandrov 3-spaces and the Maximal Volume Entropy Rigidity for $\mathsf{RCD}^*(-(N-1), N)$-spaces as well as briefly mention results in progress regarding the extension of the results of Besson, Courtois, Gallot to this setting.

2 Spaces with Curvature Bounded Below

We begin by recalling the main notations and facts of the theories of Alexandrov spaces and $\mathsf{RCD}/\mathsf{RCD}^*$ spaces. Standard references for Alexandrov Geometry are [11, 12] (see also the recent manuscript [2]). Recent and detailed references for the main aspects of $\mathsf{RCD}/\mathsf{RCD}^*$ spaces include [26] and [30].

Alexandrov spaces are part of the family of *length spaces*, which are metric spaces (X, d) such that for every $x, y \in X$, the distance can be computed by

$$d(x, y) = \inf\{L(\gamma) \mid \gamma_a = x,\ \gamma_b = y\}.$$

The infimum above is taken with respect to all continuous curves $\gamma : [a, b] \to X$, with $a \leq b$. The notation $L(\gamma)$ refers to the *length* of the curve γ. Such a length is defined by approximation by segments:

$$L(\gamma) = \sup\left\{\sum_{i=1}^{n-1} d\left(\gamma_{t_i}, \gamma_{t_{i+1}}\right)\right\},$$

where the supremum is taken over all finite partitions of $[a, b]$

$$a = t_0 \leq t_1 \leq \ldots \leq t_n = b.$$

Recall that a constant speed geodesic is a curve $\gamma : [0, 1] \to X$ such that there exists $C \geq 0$ satisfying that $d(\gamma_s, \gamma_t) = C|s - t|$ for every $s, t \in [0, 1]$. It is customary to assume that every length space X is complete and locally compact as this guarantees the existence of geodesics which realize the distance between each pair of points $x, y \in X$.

The definition of an Alexandrov space relies on the notion of *model spaces*.

For each real number k, the *model space* M_k^2 is the complete, simply-connected 2-dimensional Riemannian manifold of constant sectional curvature k. The Theorem of Killing-Hopf (see Corollary 5.6.14 in [54]) yields that M_k^2 is isometric to the standard 2-sphere, the standard Euclidean plane or the standard hyperbolic plane for $k = 1, 0, -1$.

As mentioned in the introduction the definition of an Alexandrov space is based upon triangle comparison. To clarify what this means let us recall the concept of *geodesic triangles*. A geodesic triangle Δpqr is a collection of three points $p, q, r \in X$ and three geodesics joining each pair of these points. Note that geodesics need not be unique in general so that the choice of the points does not necessarily determine the triangle. Once a geodesic triangle Δpqr in X has been fixed, one says that a geodesic triangle $\Delta \tilde{p}\tilde{q}\tilde{r}$ in M_k^2 is a *comparison triangle for* Δpqr if

$$d(p,q) = d_{M_k^2}(\tilde{p},\tilde{q}), \quad d(q,r) = d_{M_k^2}(q,r) \quad \text{and} \quad d(r,p) = d_{M_k^2}(\tilde{r},\tilde{p}).$$

Definition 2.1 A length space (X, d) has *curvature bounded below by* $k \in \mathbb{R}$ if for every $x \in X$, there is an open neighborhood $U \subset X$ of x such that for every geodesic triangle Δpqr with sides γ_{pq}, γ_{qr} and γ_{rp} and any comparison triangle $\Delta \tilde{p}\tilde{q}\tilde{r}$ with corresponding sides $\tilde{\gamma}_{\tilde{p}\tilde{q}}$, $\tilde{\gamma}_{\tilde{q}\tilde{r}}$ and $\tilde{\gamma}_{\tilde{r}\tilde{p}}$ in M_k^2 the following T_k*-property* holds true:

$d(r,s) \geq d_{M_k^2}(\tilde{r},\tilde{s})$ for every $s \in \gamma_{pq}$ and $\tilde{s} \in \tilde{\gamma}_{\tilde{p}\tilde{q}}$ such that $d(p,s) = d_{M_k^2}(\tilde{p},\tilde{s})$.

We denote this by $\mathrm{curv}(X, d) \geq k$ (or simply $\mathrm{curv}(X) \geq k$).

We can now recall the definition of an Alexandrov space.

Definition 2.2 An *Alexandrov space* is a complete and locally compact length space (X, d) with $\mathrm{curv}(X) \geq k$ for some $k \in \mathbb{R}$.

An equivalent definition can be given in terms of the so-called *monotonicity of angles condition*: Let $\alpha, \beta : [0, 1] \to X$ be geodesics such that $\alpha_0 = \beta_0$. Then we have the following characterization of an Alexandrov space: X is an Alexandrov space of curv $\geq k$ if and only if the function

$$\theta_k(s,t) := \angle \widetilde{\alpha_s}\widetilde{\alpha_0}\widetilde{\beta_t}, \tag{1}$$

where, $\Delta\widetilde{\alpha_s}\widetilde{\alpha_0}\widetilde{\beta_t}$ is a comparison triangle for $\Delta\alpha_s\alpha_0\beta_t$ is monotone non-increasing in $s, t \in [0, 1]$.

The T_k-property used to define an Alexandrov space is a local property. Nonetheless, Alexandrov spaces exhibit a local-to-global property in the sense that once the T_k-property is satisfied on all open sets of a cover of the space then it holds in the large (see Theorem 10.3.1 in [11]).

The local structure of Alexandrov spaces is in general quite different from that of a manifold. It is sufficient to take the cone over a closed Riemannian manifold with sectional curvature bounded below by 1 which is non-homeomorphic to a sphere to produce an example of an Alexandrov space with a point whose neighborhoods are non-Euclidean. However, there are some similarities with the manifolds case. It is known that the Hausdorff dimension of an Alexandrov space is either a non-negative integer or infinite. Moreover, whenever the Hausdorff dimension is finite, it coincides with the topological dimension (see for example Corollary 6.5 in [12]).

The local structure of every Alexandrov geometry can be determined via the *space of directions*. The monotonicity of angles condition helps define angles between geodesics sharing an initial point. Let X be an Alexandrov space of $\operatorname{curv} \geq k$ and assume that $\alpha, \beta : [0, 1] \to X$ are two geodesics with $p := \alpha_0 = \beta_0$. since the function $\theta_k(s, t)$ of Eq. (1) is monotone non-increasing and takes values in $[0, \pi]$, one can define the *angle between* α *and* β by

$$\angle(\alpha, \beta) := \lim_{s,t} \theta_k(s, t).$$

Such an angle is independent of k in the following sense: If there exists $l \in \mathbb{R}$ such that X, in addition to satisfying $\operatorname{curv}(X) \geq k$, satisfies $\operatorname{curv}(X) \geq l$, then

$$\lim_{s,t} \theta_k(s, t) = \lim_{s,t} \theta_l(s, t).$$

In fact, it is worth recalling here that, it follows from the definition of an Alexandrov space that if $\operatorname{curv}(X) \geq k$ for some $k \in \mathbb{R}$, then $\operatorname{curv}(X) \geq l$ for every $l \leq k$.

An equivalence relation between geodesics starting at the same point is defined by declaring two such geodesics equivalent if they make angle zero. A *geodesic direction at* $x \in X$ is an equivalence class of geodesics starting at x and the collection of all geodesic directions at the same point is a metric space when equipped with the angle as metric. The *space of directions of* X *at* x is defined as the metric completion of the space of geodesic directions at x, denoted by $\Sigma_x X$.

A tangent space at each point of an Alexandrov space (X, d) now may be defined from the notion of space of directions. There are two natural ways to make the definition at each $x \in X$. One can consider the metric cone over $\Sigma_x X$ or a blow-up of X at x, i.e. take the pointed-Gromov-Hausdorff limit of metric balls $(B(x, r_i), d_i)$ when $r_i \to 0$ (where d_i is the rescaling by $1/r_i$ of the restriction of d to the ball). It is possible to prove that both methods give rise to isometric metric spaces (see Theorem 10.9.3 in [11]). The tangent space at x is denoted by $T_x X$.

The fact that both methods of defining the tangent space give equivalent spaces implies the following structure properties of $\Sigma_x X$ (see Corollary 10.9.6 in [11]).

Theorem 2.3 *Let* X *be an Alexandrov* n*-space and* $x \in X$. *Then*

(i) Σ_x *is a compact Alexandrov* $(n-1)$*-space.*
(ii) If $n \geq 2$, *then* $\operatorname{curv}(\Sigma_x) \geq 1$.
(iii) If $n = 1$, *then* Σ_x *either consists of two points or a single point.*

The previous result allows us to define the *boundary* of an Alexandrov space. The definition is made inductively: 1-dimensional spaces are topological manifolds and therefore the boundary of such a space is defined in the usual manner. Now, the boundary of an n-dimensional Alexandrov space is defined as the subset of points $x \in X$ such that $\Sigma_x X$ has non-empty boundary. The boundary of X, denoted ∂X, is a closed subset of Hausdorff-codimension 1.

In a similar fashion to the smooth manifold case, the local topology of X around x is determined by the tangent space $T_x X$ as proved in a deep Theorem of Perelman, [50].

Theorem 2.4 (Conical Neighborhood Theorem) *Let X be an Alexandrov space and $x \in X$. Then, any sufficiently small neighborhood of x is pointed-homeomorphic to $T_x X$.*

A point x on an Alexandrov n-space X is *topologically regular* if $\Sigma_x X$ is homeomorphic to an $(n-1)$-sphere $\mathbb{S}^{n-1}$. In other case, x is called *topologically singular*. We let $S(X)$ be the subset of X consisting of topologically singular points. Then $X \setminus S(X)$ is open and dense in X (see Theorem 10.8.5 in [11]). The codimension of the subset of topologically singular points which are not boundary points is at least 3. This is a consequence of the fact that Alexandrov spaces have a canonical stratification by topological manifolds (see Theorem 10.10.1 in [11] and Structure Theorem in Section III of [50]).

The usual examples of Alexandrov spaces include the following:

(i) Toponogov's distance comparison theorem grants that complete Riemannian manifolds of sectional curvature bounded below are Alexandrov spaces (see Theorem 12.2.2 in [54]).
(ii) The boundary of an open and convex set in Euclidean space $\mathbb{R}^n$ (considered with the induced metric) is a non-negatively curved Alexandrov space (see [61]).
(iii) Let X and Y be Alexandrov spaces with curv $\geq k$ and $k \leq 0$. The Cartesian product $X \times Y$ (endowed with the usual product metric) is an Alexandrov space of curv $\geq k$. For positive curvature bounds $X \times Y$ has curv $\geq k$ if one of the spaces consist only of a single point.
(iv) Let (X, d) be a metric space with $\mathrm{diam}(X) \leq \pi$. The metric cone over X (also known as Euclidean cone) is the space $(K(X), d_K)$ obtained from $X \times [0, \infty)$ by collapsing $X \times \{0\}$ to a point. The metric d_K is defined by the cosine law, that is,

$$d_K\left((x_1, t_1), (x_2, t_2))\right) = \sqrt{t_1^2 + t_2^2 - 2t_1t_2 \cos d(x_1, x_2)}.$$

Then $K(X)$ has curv ≥ 0 if and only if X satisfies curv ≥ 1.
(v) Let (X, d) be a space with $\mathrm{diam}(X) \leq \pi$. The spherical suspension $(\mathrm{susp}(X), d_S)$ of X is the space obtained from $X \times [0, \pi]$ by collapsing $X \times \{0\}$ and $X \times \{\pi\}$ to single points. A metric d_S is then defined by the

spherical cosine law, that is:

$$\cos d_S\,((x_1,t_1),(x_2,t_2))) = \cos t_1 \cos t_2 + \sin t_1 \sin t_2 \cos d(x_1,x_2).$$

If X has curv ≥ 1, then curv susp$(X) \geq 1$.

(vi) Let $\{X_i\}_{i=1}^{\infty}$ be a sequence of compact Alexandrov spaces (not necessarily of the same dimension) with curv$(X_i) \geq k$ for all i. If X_i converges in the Gromov–Hausdorff distance to a metric space X, then X satisfies curv $\geq k$.

(vii) Let X_1 and X_2 be Alexandrov spaces with the same curvature bound curv $\geq k$, having non-empty boundaries such that ∂X_1 is isometric to ∂X_2 when considered with the induced metrics. If $f : \partial X_1 \to \partial X_2$ is an isometry, then the adjunction space $X_1 \cup_f X_2$ obtained by gluing X_1 to X_2 along their isometric boundaries is an Alexandrov space of curv $\geq k$, (see Theorem 2.1 in [56]).

We now turn to the basic aspects of Ricci curvature lower bounds for metric measure spaces. The first notion of a metric measure space having "Ricci curvature bounded below by $K \in \mathbb{R}$ and dimension bounded above by $N \in (1,\infty]$" was considered by Lott, Sturm and Villani (see [39, 64] and [65]), resulting in the class of so-called $\mathsf{CD}(K,N)$ spaces (here, CD stands for "curvature-dimension").

In order to introduce the definition let us recall a few notions from the theory of Optimal Transport. For a detailed account, we refer the reader to [3]. Let (X,d) be a complete and separable space. The set of Borel probability measures on X is denoted by $\mathcal{P}(X)$. We now consider a Borel function $c : X \times Y \to \mathbb{R} \cup \{+\infty\}$ which we refer to as the *cost function*. The problem of Optimal Transport in the formulation of Monge is the following:

Let $\mu \in \mathcal{P}(X)$ and $\nu \in \mathcal{P}(Y)$. We want to minimize the functional

$$T \mapsto \int_X c(x, T(x))\, \mathrm{d}\mu(x)$$

among all Borel maps $T : X \to Y$ such that $T_{\#}\mu = \nu$.

The previous problem is in general, either not well-posed, or no maps T satisfying the constraint may exist (see for example the discussion in the Introduction of [60]). To overcome this, a relaxation of the problem was introduced by Kantorovich as follows:

Let $\mu \in \mathcal{P}(X)$ and $\nu \in \mathcal{P}(Y)$. We want to minimize the functional

$$\gamma \mapsto \int_{X\times Y} c(x,y)\, \mathrm{d}\gamma(x)$$

among all so-called *transport plans* γ from μ to ν, that is, $\gamma \in \mathcal{P}(X \times Y)$ such that $(\pi_X)_{\#}\gamma = \mu$ and $(\pi_Y)_{\#}\gamma = \nu$, where π_X and π_Y are the canonical projections from $X \times Y$ to X and Y respectively.

The main advantage over the Monge formulation of the problem is that, under the mild assumption that the cost function is lower semi-continuous and bounded below, minimizer transport plans always exist (see Theorem 1.5 in [3]). Such a transport plan is called an *optimal transport plan*. We will let $\mathrm{Opt}(\mu, \nu)$ be the set of optimal transport plans.

Let us now consider a metric measure space (X, d, m) where (X, d) is complete and separable and m is a Borel measure. We let $\mathcal{P}_2(X, d, m)$, the family of probability measures on X with finite second moment and $\mathrm{Geo}(X)$ be the set of geodesics of X. The *Wasserstein metric* W_2 on $\mathcal{P}_2(X, d, m)$ is defined by

$$W_2^2(\mu, \nu) := \int_{X\times Y} d^2(x, y)\, \mathrm{d}\gamma(x, y)$$

where $\gamma \in \mathrm{Opt}(\mu, \nu)$. The metric space $(\mathcal{P}_2(X, d, m), W_2)$ is called the *Wasserstein space of* X, which is a complete, separable and geodesic space (see Theorems 2.7 and 2.10 in [3]).

The definition of a CD-space is inspired by a characterization in the smooth case due to Von Renesse and Sturm in terms of the geodesic (K, N)-convexity of the entropy functional on $\mathcal{P}_2(X, d, m)$, W_2 (see Theorem 1 in [66]). To proceed with the definition, let us recall the *distortion coefficients*. Given $K, N \in \mathbb{R}$ with $N \geq 0$, for $(t, \theta) \in [0, 1] \times \mathbb{R}_+$ we define

$$\sigma_{K,N}^{(t)}(\theta) := \begin{cases} \infty, & \text{if } K\theta^2 \geq N\pi^2, \\ \dfrac{\sin(t\theta\sqrt{K/N})}{\sin(\theta\sqrt{K/N})} & \text{if } 0 < K\theta^2 < N\pi^2, \\ t & \text{if } K\theta^2 < 0 \text{ and } N = 0, \text{ or if } K\theta^2 = 0, \\ \dfrac{\sinh(t\theta\sqrt{-K/N})}{\sinh(\theta\sqrt{-K/N})} & \text{if } K\theta^2 \leq 0 \text{ and } N > 0. \end{cases} \tag{2}$$

For $N \geq 1$, $K \in \mathbb{R}$ and $(t, \theta) \in [0, 1] \times \mathbb{R}_+$ we define

$$\tau_{K,N}^{(t)}(\theta) := t^{1/N} \sigma_{K,N-1}^{(t)}(\theta)^{(N-1)/N}. \tag{3}$$

For simplicity, we will provide the definition of the CD condition only in the case that the space is *essentially non-branching*. Let us recall the definition: A metric measure space (X, d, m) is essentially non-branching if for any $\mu_0, \mu_1 \in \mathcal{P}_2(X, d, m)$ which are absolutely continuous with respect to m, any optimal plan in $\mathrm{Opt}(\mu_0, \mu_1)$ is concentrated on a non-branching set of geodesics. Here, by non-branching set, we mean the following: A set $\Gamma \subset \mathrm{Geo}(X)$ is non-branching, if for every pair $\gamma, \tilde{\gamma} \in \Gamma$ such that $\gamma|_{[0,t]} = \tilde{\gamma}|_{[0,t]}$ for some $t \in (0, 1)$ then $\gamma = \tilde{\gamma}$.

With these definitions in hand we can now provide the definition of a CD-space.

Definition 2.5 (CD **Condition**) An essentially non-branching metric measure space (X, d, m) is a $\mathrm{CD}(K, N)$ space provided that for each pair $\mu_0, \mu_1 \in \mathcal{P}_2(X, d, m)$ there exists $\pi \in \mathrm{Opt}(\mu_0, \mu_1)$ such that for π-a.e. $\gamma \in \mathrm{Geo}(X)$,

$$\rho_t^{-1/N}(\gamma_t) \geq \tau_{K,N}^{(1-t)}(d(\gamma_0, \gamma_1))\rho_0^{-1/N}(\gamma_0) + \tau_{K,N}^{(t)}(d(\gamma_0, \gamma_1))\rho_1^{-1/N}(\gamma_1), \tag{4}$$

for all $t \in [0, 1]$, where $(e_t)_\sharp \pi = \rho_t m$. Here $e_t : \mathrm{Geo}(X) \to X$ is the usual evaluation map.

It is worth remarking here that on a Riemannian manifold (M^n, g), a function f of class at least C^2 is (K, N)-convex if and only if

$$\mathrm{Hess} f(v, v) \geq \frac{g(\nabla f, v)^2}{N} + K|v|^2 \qquad \text{for all } v \in TM.$$

Ohta realized that smooth compact Finsler manifolds satisfy the CD condition in [48]. To isolate the class of Riemannian-like CD-spaces, Gigli proposed in [27] to reinforce the definition of the $\mathrm{CD}(K, N)$-condition with the functional-analytic condition of *infinitesimal Hilbertianity*. To recall this notion let us introduce the concept of *minimal weak upper gradient*.

Let $C([0, 1]; X)$ be the set of continuous curves in (X, d, m). A curve $\gamma \in C([0, 1]; X)$ is said to be *absolutely continuous* provided there exists an integrable function f on $[0, 1]$ such that for every $0 \leq t < s \leq 1$,

$$d(\gamma_t, \gamma_s) \leq \int_t^s f(r)\, \mathrm{d}r.$$

Absolutely continuous curves γ have a well defined *metric speed*,

$$|\dot{\gamma}_t| := \lim_{t \to 0} \frac{d(\gamma_{t+h}, \gamma_t)}{|h|},$$

which is a function in $L^1([0, 1])$. The set of absolutely continuous curves in (X, d) is denoted by $AC([0, 1]; X)$.

A measure $\pi \in \mathcal{P}(C([0, 1]; X))$ is called a *test plan* if there exists $C > 0$ such that for every $t \in [0, 1]$,

$$(e_t)_\# \pi \leq Cm \qquad \text{and} \qquad \int\int_0^1 |\dot{\gamma}_t|^2\, \mathrm{d}t\, \mathrm{d}\pi(\gamma) < \infty.$$

The *Sobolev class* $S^2(X) := S^2(X, d, m)$ is the space of all Borel functions $f : X \to \mathbb{R}$ such that there exists a non-negative function $G \in L^2(X) := L^2(X, m)$ called *weak upper gradient* such that for any test plan π the following is satisfied

$$\int |f(\gamma_1) - f(\gamma_0)|\, \mathrm{d}\pi(\gamma) \leq \int\int_0^1 G(\gamma_t)|\dot{\gamma}_t|\, \mathrm{d}t\, \mathrm{d}\pi(\gamma).$$

It is possible to prove that there exists a minimal G, which we denote by $|\nabla f|$, called the *minimal weak upper gradient* of f.

The *Sobolev* $(1,2)$-*space* of (X, d, m) is then defined as

$$W^{1,2}(X, d, m) := L^2(X, d, m) \cap S^2(X, d, m)$$

endowed with the norm

$$\|f\|^2_{W^{1,2}(X)} := \|f\|^2_{L^2(X)} + \||\nabla f|\|^2_{L^2(X)} = \int_X (f^2 + |\nabla f|^2)\, \mathrm{d}m$$

Definition 2.6 We say that a metric measure space (X, d, m) is *infinitesimally Hilbertian* if $W^{1,2}(X)$ is a Hilbert space.

Definition 2.7 A metric measure space (X, d, m) is a $\mathsf{RCD}(K, N)$ space if it is an infinitesimally Hilbertian $\mathsf{CD}(K, N)$ space. Here RCD stands for "Riemannian-Curvature-Dimension".

At the emergence of the theory of $\mathsf{CD}(K, N)$ spaces, a central problem was to determine whether, as in the case of Alexandrov spaces, CD spaces exhibit a local-to-global property, that is, whether satisfying $\mathsf{CD}(K, N)$ for all subsets of a covering implies the condition on the full space. In fact, Rajala constructed examples of such spaces (which however are not essentially non-branching) that do not satisfy a globalization theorem (see [57]).

To address this issue, Bacher and Sturm introduced an a priori slightly weaker condition of "Ricci curvature bounded below by K with dimension at most N" known as the $\mathsf{CD}^*(K, N)$-condition, or *reduced curvature-dimension condition* (see [6]). The reduced $\mathsf{CD}^*(K, N)$ condition requires almost the same inequality as in (4) with the difference that the coefficients $\tau^{(t)}_{K,N}(\mathsf{d}(\gamma_0, \gamma_1))$ and $\tau^{(1-t)}_{K,N}(d(\gamma_0, \gamma_1))$ are replaced by $\sigma^{(t)}_{K,N}(\mathsf{d}(\gamma_0, \gamma_1))$ and $\sigma^{(1-t)}_{K,N}(d(\gamma_0, \gamma_1))$, respectively. Intuitively, the distortion coefficients of the $\mathsf{CD}(K, N)$ condition are formally obtained by imposing that one direction has linear distortion and that $N - 1$ directions are affected by curvature while the $\mathsf{CD}^*(K, N)$ condition imposes the same volume distortion in all the N directions.

It turns out that the CD and CD^* conditions are equivalent under the assumption that the space is essentially non-branching and has finite measure as shown by Cavalletti and Milman (see [14] for the result and relevant definitions).

Definition 2.8 A metric measure space (X, d, m) is a $\mathsf{RCD}^*(K, N)$ space if it is an infinitesimally Hilbertian $\mathsf{CD}^*(K, N)$ space.

As for Alexandrov spaces there are a number of constructions which produce new examples of RCD or RCD^* spaces from old ones.

(i) Any Riemannian n-manifold M regarded as a metric measure space with the induced Riemannian distance and volume density, with $\mathrm{Ric} \geq K$ is an $\mathsf{RCD}(K, n)$ space by the work of von Renesse and Sturm, [66].

(ii) For a metric measure space (X, d, m), the (K, N)-*cone* is a metric measure space defined in the following manner:
The underlying topological space $\mathrm{Con}_K(X)$ is defined as

$$\mathrm{Con}_K(X) := \begin{cases} X \times [0, \pi/\sqrt{K}]/(F \times \{0, \pi/\sqrt{K}\}) & \text{if } K > 0, \\ X \times [0, \infty)/(F \times \{0\}) & \text{if } K \leq 0. \end{cases}$$

The metric d_{Con_K} is defined, for each pair $(x, s), (y, t) \in \mathrm{Con}_K(X)$ as

$$\begin{cases} \cos^{-1}\left(\cos_K(s)\cos_K(t) + K \sin_k(s)\sin_k(t)\cos(\max\{d(x, y), \pi\})\right) & \text{if } K \neq 0, \\ \sqrt{s^2 + t^2 - 2st\cos(\max\{d(x, y, \pi\})} & \text{if } K = 0. \end{cases}$$

where

$$\sin_K(t) := \begin{cases} \frac{1}{\sqrt{K}}\sin(\sqrt{K}t) & \text{for } K > 0, \\ t & \text{for } K = 0, \\ \frac{1}{\sqrt{|K|}}\sinh(\sqrt{|K|}t) & \text{for } K < 0, \end{cases}$$

and

$$\cos_K(t) := \begin{cases} \cos(\sqrt{K}t) & \text{for } K > 0, \\ 1 & \text{for } K = 0, \\ \cosh(\sqrt{|K|}t) & \text{for } K < 0. \end{cases}$$

The measure is defined as $m^N_{\mathrm{Con}_K} := \sin^N_K(t)\, dt \otimes m$.
We observe that this generalizes the notions of Euclidean cone and Spherical suspension. It was shown by Ketterer in [36] that if (X, d, m) has diam $\leq \pi$ and satisfies $\mathsf{RCD}^*(N-1, N)$ for $N \geq 1$ then $(\mathrm{Con}_K(X), d_{\mathrm{Con}_K}, m^N_{\mathrm{Con}_K})$ satisfies $\mathsf{RCD}^*(KN, N+1)$ for $K \geq 0$.

(iii) Petrunin showed in [55] that every Alexandrov space is an RCD space.

(iv) Let $\{(X_i, d_i, m_i)\}_{i=1}^{\infty}$ be a sequence of compact metric measure spaces satisfying the $\mathsf{RCD}(K, N)$ (or $\mathsf{RCD}^*(K, N)$ condition. If (X_i, d_i, m_i) converges in the measured Gromov–Hausdorff distance to a metric measure space (X, d, m), then this limit satisfies the $\mathsf{RCD}(K, N)$ condition ($\mathsf{RCD}^*(K, N)$ condition respectively).

3 Topological Rigidity of Alexandrov 3-Spaces

In this section we will briefly survey two known instances, in the context of Alexandrov 3-spaces, in which topological rigidity holds true. Galaz-García asked in [22] whether the Borel Conjecture is still valid for Alexandrov 3-spaces. In fact, in both the instances we present, asphericity implies under appropriate assumptions that a closed aspherical Alexandrov 3-space must be homeomorphic to a 3-manifold. Therefore, the validity of the Borel Conjecture for 3-manifolds implies its validity for some classes of Alexandrov 3-spaces. Throughout this section we follow [46] and [7] closely.

Let us recall a few facts about the topology of Alexandrov 3-spaces following Galaz-García and Guijarro, [23]. Let X denote a closed Alexandrov 3-space and for each $x \in X$, let $\Sigma_x X$ be the space of directions at x. Combining Theorem 2.3 with the Bonnet-Myers Theorem (see Theorem 10.4.1. in [11]) and the classification of closed surfaces we can conclude that the homeomorphism type of $\Sigma_x X$ is that of a 2-sphere $\mathbb{S}^2$ or that of a real projective plane $\mathbb{R}P^2$. Therefore the space of directions at each topologically singular point is homeomorphic to a real projective plane. Furthermore, Theorem 2.4 coupled with the compactness of X and the codimension estimates of $S(X)$ implies that $S(X)$ is a finite set.

Topologically, a closed Alexandrov 3-space X can be described as a compact 3-manifold M having a finite and even number of $\mathbb{R}P^2$-boundary components with a cone over $\mathbb{R}P^2$ attached on each boundary component. Note that the number of $\mathbb{R}P^2$-boundary components of M is the same as the number of topologically singular points of X. The fact that this number is even can be readily seen as follows. Let DM be the double of M, that is, the closed 3-manifold obtained by gluing two copies of M along the corresponding boundary components via the identity map. One can use the Mayer-Vietoris sequence (with respect to the decomposition of DM as these two copies of M) to obtain that

$$\chi(DM) = 2\chi(M) - \chi(\partial M).$$

As DM is a closed 3-manifold, $\chi(DM) = 0$, so that the previous formula implies that $\chi(\partial M)$ is even. Moreover, as ∂M is a collection of real projective planes and $\chi(\mathbb{R}P^2) = 1$, it follows that ∂M must consist of an even number of connected components.

In the case that $S(X) \neq \emptyset$ there is an alternative topological description of X as quotient of a closed, orientable, topological 3-manifold $\tilde{X}$ by an orientation-reversing involution $\iota : \tilde{X} \to \tilde{X}$ having only isolated fixed points. The 3-manifold $\tilde{X}$ is called the *branched orientable double cover of* X (see Lemma 1.7 in [23]). The Alexandrov metric on X can be lifted to an Alexandrov metric on $\tilde{X}$ having the same lower curvature bound. Moreover, such a lifting turns ι into a global isometry. In particular, ι is equivalent to a smooth involution on $\tilde{X}$ when considered as a smooth 3-manifold. A detailed description of this construction can be found in Lemma 1.8 of [23], Sect. 2.2 of [19] and Sect. 5 of [32]; see also [33] for a higher dimensional analogue of this construction.

The first instance of topological rigidity we present occurs for closed Alexandrov 3-spaces admitting an isometric circle action. The following connected sum decomposition was proved by the author in Theorem 1.2 of [46] generalizing to the Alexandrov setting previous results of Raymond and Orlik and Raymond obtained for 3-manifolds (see [47, 58]).

Theorem 3.1 *Let X be a closed Alexandrov 3-space admitting an effective isometric S^1-action. Assume that X has $2r \geq 0$ topologically singular points. Then X is (equivariantly) homeomorphic to*

$$M\# \underbrace{\mathrm{susp}(\mathbb{R}P^2)\# \cdots \# \mathrm{susp}(\mathbb{R}P^2)}_{r \text{ summands}},$$

where M is a closed 3-manifold with a S^1-action.

From this result, first one can show that a connected sum

$$Y := \mathrm{Susp}(\mathbb{R}P^2)\#\mathrm{Susp}(\mathbb{R}P^2)$$

is not aspherical. Since the argument is simple we outline it here.

One shows via Van Kampen's Theorem that Y is simply connected so that the Hurewicz theorem can be applied. Then we obtain that $\pi_2(Y) \cong H_2(Y)$. Then one shows via the exact sequence of the pair $(Y, \mathbb{S}^2)$ (where $\mathbb{S}^2$ is the sphere used in the connected sum construction) that there is a surjection

$$H_2(Y) \to H_2(\mathrm{Susp}(\mathbb{R}P^2) \vee \mathrm{Susp}(\mathbb{R}P^2))$$

Here, $\mathrm{Susp}(\mathbb{R}P^2) \vee \mathrm{Susp}(\mathbb{R}P^2)$ denotes the wedge of $\mathrm{Susp}(\mathbb{R}P^2)$ with itself, where the distinguished point used in the construction is a topologically regular point. Then we have that $H_2(\mathrm{Susp}(\mathbb{R}P^2) \vee \mathrm{Susp}(\mathbb{R}P^2)) \neq 0$ by the Mayer-Vietoris sequence. Whence $H_2(Y) \neq 0$, proving the claim.

Now, we can show that a connected sum of the form $M\#Z$, where M is a closed 3-manifold and $Z := \mathrm{susp}(\mathbb{R}P^2)$ is not aspherical. One proceeds by contradiction as follows. Assume that $M\#Z$ is aspherical and consider the map $\varphi : M\#Z \to M \vee Z$ that collapses the $\mathbb{S}^2$ used in the connected sum construction to a point. One can lift φ to a 2-connected map between universal covers $\tilde{\varphi} : \widetilde{M\#Z} \to \widetilde{M \vee Z}$. Then, by the asphericity of $M\#Z$ and the Hurewicz Theorem we have that $H_2(\widetilde{M \vee Z}) = 0$. We now let p denote the universal cover map $\widetilde{M \vee Z} \to M \vee Z$. Observe that $p^{-1}(M) \cap p^{-1}(Z) = p^{-1}(\{pt\})$ is a discrete set and that $p^{-1}(M) \cup p^{-1}(Z) = \widetilde{M \vee Z}$. It now follows from the Mayer-Vietoris sequence that $H_2(p^{-1}(M)) \oplus H_2(p^{-1}(Z)) \cong 0$. The preimage of Z is a disjoint union of copies of the universal cover of Z and since

Z is simply connected, $p^{-1}(Z)$ is a disjoint union of copies of Z. Hence, $H_2(Z) = 0$ which is a contradiction. Therefore we obtain the following result:

Theorem 3.2 *Let X be a closed Alexandrov* 3*-space admitting an effective and isometric circle action. If X is aspherical, then X is homeomorphic to a closed* 3*-manifold.*

Corollary 3.3 *Let X and Y be closed Alexandrov* 3*-spaces admitting effective and isometric circle actions. If X is homotopy equivalent to Y, then X is homeomorphic to Y.*

Group actions can be thought of as a particular case of the following phenomenon known as *collapse*. Given a sequence $\{X_i\}_{i=1}^{\infty}$ closed Alexandrov 3-spaces with diameters uniformly bounded above by $D > 0$ and $\operatorname{curv} X_i \geq k$ for some $k \in \mathbb{R}$, Gromov's Precompactness Theorem implies that (possibly after passing to a subsequence), there exists an Alexandrov space Y with diameter bounded above by D and $\operatorname{curv} Y \geq k$ such that $X_i \xrightarrow{GH} Y$. In the case in which $\dim Y < 3$, the sequence X_i is said to *collapse to* Y. Similarly, a closed Alexandrov 3-space X is a *collapsing Alexandrov* 3*-space* if there exists a sequence of Alexandrov metrics d_i on X such that the sequence $\{(X, d_i)\}_{i=1}^{\infty}$ is a collapsing sequence.

The topological classification of closed collapsing Alexandrov 3-spaces was carried out by Mitsuishi and Yamaguchi in [43]. We now give a brief summary of the classification.

The classification of the topology of X_i for $\dim Y = 2$ is contained in Theorems 1.3 and 1.5 of [43]; In the case in which $\partial Y = \emptyset$, X_i is homeomorphic to a *generalized Seifert fibered space* $\operatorname{Seif}(Y)$ (see Definition 2.48 [43]) for sufficiently big i.

In the case in which $\partial Y \neq \emptyset$, $\operatorname{Seif}(Y)$ is attached with a finite number of *generalized solid tori and Klein bottles* (see Definition 1.4 of [43]).

For $\dim Y = 1$ and $\partial Y = \emptyset$, X_i is homeomorphic to an fiber bundle over $\mathbb{S}^1$ with fiber F, where F is homeomorphic to a torus T^2, a Klein bottle K^2, a 2-sphere $\mathbb{S}^2$ or a real projective plane $\mathbb{R}P^2$, for big enough i (see Theorem 1.7 in [43]).

On the other hand, by Theorem 1.8 in [43], if $\partial Y \neq \emptyset$, X_i is homeomorphic to a union of two spaces B and B' with one boundary component, glued along their homeomorphic boundaries, where ∂B is one of the spaces T^2, K^2, $\mathbb{S}^2$ or $\mathbb{R}P^2$. The pieces B and B' are the following:

(i) If $\partial B \cong \mathbb{S}^2$ then B and B' are homeomorphic to one of: a 3-ball D^3, a 3-dimensional projective space with the interior of a 3-ball removed $\mathbb{R}P^3 \setminus \operatorname{int} D^3$ or $\mathrm{B}(S_2)$, a space homeomorphic to a small metric ball of an $\mathbb{S}^2$-soul of an open non-negatively curved Alexandrov space.
(ii) If $\partial B \cong \mathbb{R}P^2$ then B and B' are homeomorphic to a closed cone over a projective plane.
(iii) If $\partial B \cong T^2$ then B and B' are homeomorphic to $\mathbb{S}^1 \times D^2$, $\mathbb{S}^1 \times \mathrm{Mo}$ (where Mo is a Mobius strip), the orientable non-trivial I-bundle over K^2, $K^2 \tilde{\times} I$ or

B(S_4), a space homeomorphic to a small metric ball of an $\mathbb{S}^2$-soul of an open non-negatively curved Alexandrov space.

(iv) If $\partial B \cong K^2$ then B and B' are homeomorphic to $\mathbb{S}^1 \tilde{\times} D^2$ the non-orientable D^2-bundle over $\mathbb{S}^1$, $K^2 \hat{\times} I$ the non-orientable non-trivial I-bundle over K^2, the space B(pt) defined in Example 1.2 of [43], or B$(\mathbb{R}P^2)$, a space homeomorphic to a small metric ball of an $\mathbb{R}P^2$-soul of an open non-negatively curved Alexandrov space.

Finally, if $\dim Y = 0$, Theorem 1.9 of [43] states that, for i sufficiently big, X_i is homeomorphic to either a generalized Seifert fiber space Seif(Z), (where Z is a 2-dimensional Alexandrov space with $\operatorname{curv} Z \geq 0$), a space appearing in the cases in which $\dim Y = 1, 2$ or a non-negatively curved Alexandrov space with finite fundamental group.

To present another instance of topological rigidity in the context of collapsing Alexandrov 3-spaces, recall that a closed Alexandrov 3-space X is *irreducible* if every embedded 2-sphere in X bounds a 3-ball and, in the case that $S(X) \neq \emptyset$, it is further required that every 2-sided $\mathbb{R}P^2$ bound a cone over the real projective plane $K(\mathbb{R}P^2)$. The following result was proved by Bárcenas and the author in [7], showing that irreducible and "sufficiently collapsed" Alexandrov 3-spaces which are aspherical are in fact homeomorphic to 3-manifolds.

Theorem 3.4 *For any $D > 0$, there exists $\varepsilon = \varepsilon(D) > 0$ such that, if X is irreducible and aspherical and* $\operatorname{diam} X \leq D$ *and* vol $\leq \varepsilon$ *then X is homeomorphic to a 3-manifold.*

The proof is based on the classification of Mitusishi and Yamaguchi and an analysis of asphericity of the spaces and their branched orientable double covers on a case by case basis depending on the dimension of the limit space. The classification of local circle actions on Alexandrov 3-spaces by Galaz-García and the author (see [25]) also plays an important role. The strategy of the proof closely resembles that of Theorem A in [24]. Theorem 3.4 directly implies the following Corollary asserting the validity of the Borel conjecture for irreducible and sufficiently collapsed Alexandrov 3-spaces.

Corollary 3.5 *For any $D > 0$, there exists $\varepsilon = \varepsilon(D) > 0$ such that, if X and Y are aspherical and irreducible and have* diam $\leq$ *D and volumes $\leq \varepsilon$, then the Borel Conjecture holds for X and Y, that is, if X is homotopy equivalent to Y then X is homeomorphic to Y.*

Theorems 3.1 and 3.4 seem to suggest that asphericity may be a strong enough property to prevent the existence of topologically singular points in every Alexandrov 3-space. This conjecture was proposed in [7]. It is worth noting that the conjecture is true for simply-connected, closed, aspherical Alexandrov 3-spaces (see Remark 3.5 in [7]).

Conjecture 3.6 Every closed, aspherical Alexandrov 3-space is homeomorphic to a 3-manifold.

4 Volume Entropy Rigidity of RCD^*-Spaces

Volume entropy is a fundamental geometric invariant, related to several other quantities, such as the topological entropy of geodesic flows, minimal volume, simplicial volume, bottom spectrum of the Laplacian of the universal cover, among others. The *volume entropy* of a compact Riemannian manifold (M^n, g), measures the exponential growth rate of the volume of balls in the universal cover. It is defined as

$$h(M, g) = \lim_{R\to\infty} \frac{\log(\mathrm{vol}(B(\tilde{x}, R)))}{R},$$

where $B(\tilde{x}, R)$ is a ball in the universal cover $\tilde{M}$ of M. It can be shown that the limit exists and is independent of the base point $\tilde{x} \in \tilde{M}$ (see [41]).

When $\mathrm{Ric} \geq -(n-1)$, the Bishop-Gromov volume comparison gives the upper bound $h(M, g) \leq n-1$, which is the volume entropy of any hyperbolic n-manifold. As we mentioned in the Introduction, the equality case was analyzed by Ledrappier and Wang in [37]. Liu found a simpler proof in [38], and recently Chen, Rong and Xu obtained the following quantitative version of this rigidity result in [16].

Theorem 4.1 *Given n, $D > 0$, there exists $\varepsilon(n, d) > 0$ such that for any $0 < \varepsilon < \varepsilon(n, d)$ if a compact, Riemannian n-manifold M satisfies $\mathrm{Ric} \geq -(n-1)$, $\mathrm{diam} \leq D$ and $h(M, g) \geq n - 1 - \varepsilon$, then M is diffeomorphic to a hyperbolic manifold by a $\Psi(\varepsilon, n, D)$-isometry, where $\Psi \to 0$ as $\varepsilon \to 0$.*

The same type of maximal volume entropy rigidity can be shown to hold for $\mathsf{RCD}^*(K, N)$-spaces as was carried out by Connell, Dai, Perales, Suárez-Serrato, Wei and the author in [18]. Mondino and Wei showed that the universal cover of an $\mathsf{RCD}^*(K, N)$ space with $1 < N < \infty$ exists and is also an $\mathsf{RCD}^*(K, N)$ space, [44]. It is worth observing however that the universal cover is meant in the categorical sense and may not be simply-connected. This allows us to define the volume entropy similarly for a compact $\mathsf{RCD}^*(K, N)$ space (X, d, m), that is, let (X, d, m) be a compact $\mathsf{RCD}^*(K, N)$ space, and $(\widetilde{X}, \tilde{d}, \tilde{m})$ its universal cover. We define the *volume growth entropy* of (X, d, m) as

$$h(X, d, m) := \lim_{R\to\infty} \frac{1}{R} \log\left(\widetilde{m}(B_{\widetilde{X}}(x, R))\right).$$

The volume growth entropy is well defined, and it is independent of x and the measure m, (see [59] and [10]). Observe that if (M, g) is a Riemannian manifold then with the induced Riemannian distance and the volume measure $m = \mathrm{d\,vol}$, then this definition coincides with the usual definition.

Theorem 4.2 *Let $1 < N < \infty$ and (X, d, m) be a compact $\mathsf{RCD}^*(-(N-1), N)$ space. Then $h(X) \leq N - 1$, and the equality holds if and only if N is an integer and the universal cover of X is isometric to the N-dimensional real hyperbolic space via a measure preserving isometry (up to rescaling of the measure by a constant factor).*

The analogous result for Alexandrov spaces was shown by Jiang in [34] using different techniques.

Rigidity results for RCD^* spaces often imply stability results directly as RCD^* spaces are closed under measured Gromov-Hausdorff convergence. Theorem 4.2 implies the following quantitative version. Observe that we assume a lower bound on the first systole as this ensures the continuity of the volume entropy under measured Gromov-Hausdorff convergence (see [59]).

Theorem 4.3 *Let* $1 < N < \infty$, $v > 0$, $D > 0$. *There exists* $\epsilon(N, v, D) > 0$ *such that for* $0 < \epsilon < \epsilon(N, v, D)$, *if* (X, d, m) *is a compact* $\mathsf{RCD}^*(-(N-1), N)$ *space satisfying*

(i) $\mathrm{diam}(X) \leq D$,
(ii) $h(X) \geq N - 1 - \epsilon$,
(iii) $\mathrm{sys} \geq L$,

then X *is* $\Psi(\epsilon, N, L, D)$ *measured Gromov-Hausdorff close to an* N*-dimensional hyperbolic manifold.*

It is conjectured that the uniform lower bound in the volume is not necessary. In fact when X is a manifold, a diameter upper bound and almost maximal volume entropy imply non-collapsing as proved in [16].

The upper bound for the volume entropy in Theorem 4.2 is obtained as in the smooth case, via the Bishop-Gromov volume comparison theorem, also available for RCD^*-spaces (see Theorem 1.1 in [15] and Theorem 5.1 [49]).

The strategy and techniques for the proof of the equality case in Theorem 4.2 resemble those of the Splitting Theorem for $\mathsf{RCD}(0, N)$-spaces due to Gigli, [28]. To explain the main ideas in the proof, let us recall a few notions.

The domain of the Laplacian $D(\Delta)$ is defined in the distributions sense, as the subset of $W^{1,2}(X, d, m)$ of all g such that for some $h \in L^2(X, m)$,

$$- \int \langle \nabla f, \nabla g \rangle \, \mathrm{d}m = \int f h \, \mathrm{d}m \tag{1}$$

for all $f \in W^{1,2}(X)$. The function h is denoted as $\Delta g = h$. We can similarly define a local version of the domain of the Laplacian, $D_{loc}(\Delta)$, to be the corresponding subset of $W^{1,2}_{loc}(X) := L^2_{loc}(X, m) \cap S^2_{loc}(X, d, m)$, namely the subset of all $g \in W^{1,2}_{loc}(X)$ such that the Eq. (1) holds for all $f \in \mathrm{Test}_{bs}(X)$. Here,

$$\mathrm{Test}(X) := \left\{ f \in D(\Delta) \cap L^\infty(X, m) \mid |\nabla f| \in L^\infty(X, m) \text{ and } \Delta f \in W^{1,2}(X) \right\},$$

and $\mathrm{Test}_{bs}(X)$ is the subset of $\mathrm{Test}(X)$ consisting of functions with bounded support.

In order to show that the universal cover $(\tilde{X}, \tilde{d}, \tilde{m})$ of (X, d, m) is isometric via a measure preserving isometry to a hyperbolic space, one shows that $\tilde{X}$ is isometric to a warped product space of the form $X' \times_{e^t} \mathbb{R}$ (again, via a measure preserving isometry), and then show that X' is regular enough. At this point the analogy with

[28] becomes clear, as now the problem can be regarded as a warped-product version of the splitting theorem under the assumption of maximality of volume entropy.

To obtain a metric measure space which is a candidate for the role of X', one constructs a Busemann-type function $u : \tilde{X} \to \mathbb{R}$ in $D_{loc}(\Delta)$ with $|\nabla u| = 1$ and $\Delta u = N - 1$ by taking advantage of the maximality of the volume entropy and a covering argument due to Liu, [38]. Then one proceeds to show this function admits a so-called *Regular Lagrangian Flow*.

A Regular Lagrangian Flow is an analogue of a gradient flow defined by Ambrosio and Trevisan in [5] which exists in very low regularity settings. To recall the definition, let us first recall the concept of tangent module.

For an infinitesimally Hilbertian metric measure space (X, d, m) there exists a unique couple $(L^2(T^*X), d)$ (up to isomorphism of modules) where $L^2(T^*X)$ is an $L^2(m)$*-normed* $L^\infty(m)$*-module* (see Definition 1.2.10 in [29]) and $d : S^2(X) \to L^2(T^*X)$ is a linear operator such that the following two conditions hold

(i) $|df| = |\nabla f|$ m-a.e. for every $f \in S^2(X)$. Here $|df|$ denotes the pointwise norm of df in $L^2(T^*X)$.
(ii) $L^2(T^*X)$ is spanned by $\{df \mid f \in S^2(X)\}$.

The module $L^2(T^*X)$ is called the *cotangent module of* X and d is the *differential*. The tangent module of X, denoted by $L^2(TX)$ is defined as the dual module of $L^2(T^*X)$ and the *gradient* $\nabla f \in L^2(TX)$ of a function $f \in W^{1,2}(X)$ is the unique element associated to df via the Riesz isomorphism. Intuitively, $L^2(TX)$ plays the role of the "space of L^2 sections of the tangent bundle", though note that there is no actual tangent bundle defined in general. A version with local integrability conditions can be defined as follows: $L^2_{loc}(TX)$ is the set of "vector fields" V such that $|V| \in L^2_{loc}(X, m)$.

We now recall the definition of a Regular Lagrangian Flow.

Definition 4.4 Let $V \in L^2_{loc}(TX)$. We say that

$$F^V : [0, 1] \times X \to X$$

is a *Regular Lagrangian Flow* for V provided that:

(i) There exists $C > 0$ such that

$$(F_s^V)_{\#}m \leq Cm, \quad \forall s \in [0, 1].$$

(ii) For m-a.e. $x \in X$ the curve $[0, 1] \ni s \mapsto F_s^V(x) \in X$ is continuous and such that

$$F_0^V(x) = x.$$

(iii) For every $f \in W^{1,2}(X)$ we have that for m-a.e. $x \in X$ the function $s \mapsto f(F_s^V(x))$ belongs to $W^{1,1}(0,1)$ and satisfies

$$\frac{d}{ds} f(F_s^V(x)) = df(V)(F_s^V(x)), \quad m \times \mathcal{L}^1|_{[0,1]} - a.e.(x,s).$$

The trajectories $F_{(\cdot)}(x)$ induce a partition of $\tilde{X}$ (up to a set of measure zero) and the high regularity of u provides information on how the reference measure $\tilde{m}$ and the distance change under the action of the flow. Therefore, the natural candidate for X' is $u^{-1}(0)$, the slice at time 0 of the partition induced by F, equipped with the natural intrinsic metric and an appropriately defined measure which agrees with the data provided by F.

At this point we need to show that the natural maps from and into $\tilde{X}$ and $\mathbb{R} \times_{e^t} X'$ are isomorphisms (that is, measure-preserving isometries) of metric measure spaces. This can be shown at the level of Sobolev spaces making use of the structure of Sobolev spaces for warped products studied by Gigli and Han in [31]. Once an isometry is obtained for the Sobolev spaces it is possible to obtain the isomorphism at the level of the metric measure spaces using a deep result of Gigli which we recall below (see Proposition 4.20 in [28]).

Definition 4.5 A metric measure space (X, d, m) has the *Sobolev to Lipschitz property* if any $f \in W^{1,2}(X, d, m)$ with $|\nabla f| \leq 1$ m-a.e. admits a 1-Lipschitz representative, i.e., a 1-Lipschitz function $g : X \to \mathbb{R}$ such that $f = g$ m-a.e..

Gigli showed that $\mathsf{CD}(K, N)$-spaces have the Sobolev-to-Lipschitz property and Ambrosio, Gigli and Savaré showed this fact for $\mathsf{RCD}(K, \infty)$-spaces. Since $\mathsf{CD}^*(K, N)$ spaces are $\mathsf{CD}(K^*, N)$ spaces for a suitable value of K^*, $\mathsf{RCD}^*(K, N)$ spaces also satisfy the Sobolev-to-Lipschitz property (see [13] and [15]).

Theorem 4.6 (Isomorphisms via Duality with Sobolev Norms) *Let (X_1, d_1, m_1) and (X_2, d_2, m_2) be two metric measure spaces with the Sobolev-to-Lipschitz property and $T : X_1 \to X_2$ a Borel map. Assume that both m_1 and m_2 give finite mass to bounded sets. Then the following are equivalent.*

(i) Up to a modification on a m_1-negligible set, T is an isomorphism of the metric measure spaces
(ii) The following two statements are true.

(iia) There exist a Borel m_1-negligible set $\mathcal{N} \subset X_1$ and a Borel map $S : X_2 \to X_1$ such that

$$S(T(x)) = x, \quad \textit{for all} \quad x \in X_1 \setminus \mathcal{N}.$$

(iib) *The right composition with T produces a bijective isometry of $W^{1,2}(X_2, d_2, m_2)$ in $W^{1,2}(X_1, d_1, m_1)$, i.e. $f \in W^{1,2}(X_2, d_2, m_2)$ if and only if $f \circ T \in W^{1,2}(X_1, d_1, m_1)$ and in this case*

$$\|f\|_{W^{1,2}(X_2)} = \|f \circ T\|_{W^{1,2}(X_1)}.$$

To proceed, we recall that the definition of an $\mathsf{RCD}(K, N)$ space is equivalent to the validity of the following weak Bocnher's inequality by the work of Ambrosio, Mondino and Savaré in [4] and Erbar, Kuwada and Sturm in [20].

Theorem 4.7 (Weak Bochner's Inequality) *Let (X, d, m) be an $\mathsf{RCD}^*(K, N)$-space. Then, for all $f \in \mathbf{D}(\Delta)$ with $\Delta f \in W^{1,2}(X, d, m)$ and all $g \in \mathbf{D}(\Delta) \cap L^\infty(X, m)$ non-negative with $\Delta g \in L^\infty(X.m)$ we have*

$$\frac{1}{2}\int \Delta g|\nabla f|^2 dm - \int g\,\langle\nabla(\Delta f), \nabla f\rangle\, dm \geq K\int g|\nabla f|^2 dm + \frac{1}{N}\int g(\Delta f)^2 dm.$$

The structure of a warped product space naturally implies via Bochner's inequality that X' is an $\mathsf{RCD}^*(-(N-1), N-1)$ space. To complete the proof one now applies an argument of Chen et al. [16], where one uses the fact that the entropy is maximal to show that X' must be isometric to $\mathbb{R}^{N-1}$, which shows that $\mathbb{R} \times_{e^t} X'$ is isomorphic to the N-dimensional hyperbolic space.

In the forthcoming work [17], Connell, Dai, Perales, Suárez-Serrato, Wei and the author have extended the techniques used in the work of Besson, Courtois and Gallot, [8, 9], to the context of RCD spaces, generalizing Theorem 1.4. For simplicity we state here only the case in which f is a homotopy equivalence. Furthermore, to avoid technical details we refer the reader to [35] for the definition of the (reduced) boundary of an $\mathsf{RCD}(K, N)$ space for which the reference measure is the N-dimensional Hausdorff measure.

Theorem 4.8 *Let $(X, d, \mathcal{H}^N)$ be an $\mathsf{RCD}(-(N-1), N)$ space without boundary and M_{hyp} a closed hyperbolic N-manifold of constant curvature -1. If X and M_{hyp} are homotopy equivalent, then*

$$\mathcal{H}^N(X) \geq \mathrm{vol}(M_{hyp}).$$

Moreover, equality occurs if and only if X is isometric to M_{hyp}.

Acknowledgments JNZ is supported by a DGAPA-UNAM postdoctoral scholarship. This manuscript is partially based on the talk given by the author at the 11th mini-meeting in Differential Geometry at the Centro de Investigación en Matemáticas (CIMAT). He wishes to thank the organizers for the opportunity to submit this expository article and Chris Connell for very useful communications. The author further wishes to thank the anonymous reviewer for a careful reading of the manuscript.

References

1. J. W. Alexander, *Note on two three dimensional manifolds with the same group*, Trans. Amer. Math. Soc. 20 (1919), 339–342.
2. S. Alexander, V. Kapovitch and, A. Petrunin, *Alexandrov geometry: preliminary version no. 1*, preprint `arXiv:1903.08539`.
3. L. Ambrosio and N. Gigli, *A user's guide to optimal transport*, Modelling and optimisation of flows on networks, 1–155, Lecture Notes in Math., 2062, Fond. CIME/CIME Found. Subser., Springer, Heidelberg, 2013
4. L. Ambrosio, A. Mondino and, G. Savare, *Nonlinear diffusion equations and curvature conditions in metric measure spaces*, Mem. Amer. Math. Soc. 262 (2019), no. 1270.
5. L. Ambrosio and D. Trevisan, *Well-posedness of Lagrangian flows and continuity equations in metric measure spaces*, Anal. PDE 7 (2014), no. 5, 1179–1234.
6. K. Bacher, T. Sturm, *Localization and tensorization properties of the curvature-dimension condition for metric measure spaces*, J. Funct. Anal. 259 (2010) 28–56.
7. N. Bárcenas and J. Núñez-Zimbrón, *On topological rigidity of Alexandrov 3-spaces*, Rev. Mat. Iberoam. 37 (2021), no. 5, 1629–1639.
8. G. Besson, G. Courtois and S. Gallot, *Entropies et rigidites des espaces localement symetriques de courbure strictement negative*, G.A.F.A. 5(5) (1995), 731–799.
9. ——, *Minimal entropy and Mostow's rigidity theorems*, Ergodic Theory Dynam. Systems 16 (1996), no. 4, 623–649.
10. G. Besson, G. Courtois, S. Gallot and, A. Sambusetti, *Curvature-free Margulis lemma for Gromov-hyperbolic spaces*, preprint `arXiv:1712:08386`.
11. D. Burago, Y. Burago, and, S. Ivanov, *A course in metric geometry*, Graduate Studies in Mathematics, 33. American Mathematical Society, Providence, RI, 2001.
12. Y. Burago, M. Gromov, and G. Perelman, *A. D. Aleksandrov spaces with curvatures bounded below*, (Russian) ; translated from Uspekhi Mat. Nauk 47 (1992), no. 2(284), 3–51, 222 Russian Math. Surveys 47 (1992), no. 2, 1–58.
13. F. Cavalletti, *Decomposition of geodesics in the Wasserstein space and the globalization problem*, Geom. Funct. Anal. 24 (2014), no. 2, 493–551.
14. F. Cavalletti, E. Milman, *The globalization theorem for the curvature-dimension condition*, Invent. Math. 226 (2021), no. 1, 1–137.
15. F. Cavalletti and K. Sturm, *Local curvature-dimension condition implies measure-contraction property*, J. Funct. Anal. 262 (2012), no. 12, 5110–5127.
16. L. Chen, X. Rong and S. Xu, *Quantitative volume rigidity of space form with lower Ricci curvature bound*, J. Differential Geom. 113 (2019), 227–272.
17. C. Connell, X. Dai, J. Núñez-Zimbrón, R. Perales, P. Suárez-Serrato, G. Wei, *Volume entropy and rigidity for* RCD*-spaces*, in preparation.
18. ——, *Maximal volume entropy rigidity for* $\mathsf{RCD}^*(-(N-1), N)$ *spaces*, J. Lond. Math. Soc. (2) 104 (2021), no. 4, 1615–1681.
19. Q. Deng, F. Galaz-García, L. Guijarro and M. Munn, *Three-Dimensional Alexandrov Spaces with Positive or Nonnegative Ricci Curvature*, Potential Anal. 48 (2018), no. 2, 223–238.
20. M. Erbar, K. Kuwada and, K. Sturm, *On the equivalence of the entropic curvature-dimension condition and Bochner's inequality on metric measure spaces*, Invent. Math. (2015) Vol. 201(3) 993–1071.
21. M. H. Freedman, *The topology of four-dimensional manifolds*, J. Differential Geometry 17 (1982), no. 3, 357–453.
22. F. Galaz-García, *A glance at three-dimensional Alexandrov spaces*, Front. Math. China 11 (2016), no. 5, 1189–1206.
23. F. Galaz-García and L. Guijarro, *On three-dimensional Alexandrov spaces*, Int. Math. Res. Not. IMRN 2015, no. 14, 5560–5576.

24. F. Galaz-García, L. Guijarro, and J. Núñez-Zimbrón, *Sufficiently collapsed irreducible Alexandrov 3-spaces are geometric*, Indiana Univ. Math. J., 69 (2020), no. 3, 977–1005.
25. F. Galaz-García and J. Núñez-Zimbrón, *Three-dimensional Alexandrov spaces with local isometric circle actions*, Kyoto J. Math., 60 (2020), no. 3, 801–823.
26. N. Gigli, *Nonsmooth differential geometry—an approach tailored for spaces with Ricci curvature bounded from below*, Mem. Amer. Math. Soc. 251 (2018), no. 1196.
27. ——, *On the differential structure of metric measure spaces and applications*, Mem. Amer. Math. Soc. 236 (2015), no. 1113, vi+91 pp.
28. ——, *The splitting theorem in non-smooth context*, preprint arXiv:1302.5555.
29. ——, *Nonsmooth differential geometry–an approach tailored for spaces with Ricci curvature bounded from below*, Mem. Amer. Math. Soc. 251 (2018), no. 1196, v+161 pp.
30. N. Gigli and E. Pasqualetto, *Lectures on Nonsmooth Differential Geometry*, SISSA Springer Series, 2, Springer International Publishing, 2020.
31. N. Gigli and B. Han, *Sobolev spaces on warped products*, Journal of Functional Analysis, Vol. 275, 8 (2018) 2059–2095.
32. K. Grove and B. Wilking, *A knot characterization and 1-connected nonnegatively curved 4-manifolds with circle symmetry*, Geom. Topol. 18 (2014), no. 5, 3091–3110.
33. J. Harvey and C. Searle, *Orientation and symmetries of Alexandrov spaces with applications in positive curvature*, J. Geom. Anal. 27 (2017), no. 2, 1636–1666.
34. Y. Jiang, *Maximal bottom of spectrum or volume entropy rigidity in Alexandrov geometry*, Math. Z. 291 (2019), no 1–2, 55–84.
35. V. Kapovitch, A. Mondino, *On the topology and the boundary of N-dimensional* $\mathsf{RCD}(K, N)$ *spaces*, arXiv preprint arXiv:1907.02614 (2019).
36. C. Ketterer, *Cones over metric measure spaces and the maximal diameter theorem*, J. Math. Pures Appl. (9) 103 (2015), no. 5, 1228–1275.
37. F. Ledrappier, X. Wang, *An integral formula for the volume entropy with applications to rigidity*, J. Differential Geom. 85 (2010), no. 3, 461–477.
38. G. Liu, *A short proof to the rigidity of volume entropy*, Math. Res. Lett. 18 (2011), no. 1, 151–153.
39. J. Lott, C. Villani, *Ricci curvature for metric-measure spaces via optimal transport*, Ann. of Math. (2) 169 (2009), no. 3, 903–991.
40. W. Lück, *Survey on aspherical manifolds*, European Congress of Mathematics, 53–82, Eur. Math. Soc., Zürich, 2010.
41. A. Manning, *Topological entropy for geodesic flows*, Ann. of Math. (2) 110 (1979), no. 3, 567–573.
42. J. Milnor, *On manifolds homeomorphic to the 7-sphere*, Ann. of Math. (2) 64 (1956), 399–405.
43. A. Mitsuishi and T. Yamaguchi, *Collapsing three-dimensional closed Alexandrov spaces with a lower curvature bound*, Trans. Amer. Math. Soc. 367 (2015), no. 4, 2339–2410.
44. A. Mondino, G. Wei, *On the universal cover and the fundamental group of an $RCD^*(K, N)$-space*, J. Reine Angew. Math. 753, (2019) 211–237.
45. G. D. Mostow, *Quasi-conformal mappings in n-space and the rigidity of the hyperbolic space forms*, Publ. Math. IHES, 34: 53–104
46. J. Núñez-Zimbrón, *Closed three-dimensional Alexandrov spaces with isometric circle actions*, Tohoku Math. J. (2) 70 (2018), no. 2, 267–284.
47. P. Orlik and F. Raymond, *Actions of* SO(2) *on 3-manifolds*, 1968 Proc. Conf. on Transformation Groups (New Orleans, La., 1967) pp. 297–318 Springer, New York
48. S. Ohta, *Finsler interpolation inequalities*, Calc. Var. Partial Differential Equations 36 (2009), no.2, 211–249.
49. ——, *On the measure contraction property of metric measure spaces*, Comment. Math. Helv. 82 (2007), no. 4, 805–828.
50. G. Perelman, *Elements of Morse theory on Aleksandrov spaces*, (Russian) ; translated from Algebra i Analiz 5 (1993), no. 1, 232–241 St. Petersburg Math. J. 5 (1994), no. 1, 205–213.
51. G. Perelman, *The entropy formula for the Ricci flow and its geometric applications*, preprint, arXiv:math/0211159.

52. G. Perelman, *Ricci flow with surgery on three-manifolds*, preprint, `arXiv:math/0303109`.
53. G. Perelman, *Finite extinction time for the solutions to the Ricci flow on certain three-manifolds*, preprint, `arXiv:math/0307245`.
54. P. Petersen, *Riemannian geometry*, Third edition. Graduate Texts in Mathematics, 171. Springer, Cham, 2016.
55. A. Petrunin, *Alexandrov meets Lott-Villani-Sturm*, Münster J. Math. 4 (2011), 53–64.
56. A. Petrunin, *Applications of quasigeodesics and gradient curves*, Comparison geometry (Berkeley, CA, 1993–94), 203–219, Math. Sci. Res. Inst. Publ., 30, Cambridge Univ. Press, Cambridge, 1997.
57. T. Rajala, *Failure of the local-to-global property for $CD(K, N)$ spaces*, Ann. Sc. Norm. Super. Pisa Cl. Sci. (5) 15 (2016), 45–68.
58. F. Raymond, *Classification of the actions of the circle on* 3*-manifolds*, Trans. Amer. Math. Soc. 131 (1968), 51–78.
59. G. Reviron, *Rigidité topologique sous l'hypothèse entropie majorée et applications*, Comment. Math. Helv. 83 (2008), 815–846.
60. F. Santambrogio, *Optimal transport for applied mathematicians*, Calculus of variations, PDEs, and modeling. Progress in Nonlinear Differential Equations and their Applications, 87. Birkhäuser/Springer, Cham, 2015.
61. K. Shiohama, *An introduction to the geometry of Alexandrov spaces*, Lecture Notes Series, 8. Seoul National University, Research Institute of Mathematics, Global Analysis Research Center, Seoul, 1993.
62. S. Smale, *Generalized Poincaré's conjecture in dimensions greater than four*, Ann. of Math. (2) 74 (1961), 391–406.
63. S. Smale, *The generalized Poincaré conjecture in higher dimensions*, Bull. Amer. Math. Soc. 66 (1960), 373–375.
64. K. T. Sturm, *On the geometry of metric measure spaces. I*, Acta Math. 196 (2006), no. 1, 65–131
65. ——, *On the geometry of metric measure spaces. II*, Acta Math. 196 (2006), no. 1, 133–177.
66. M.-K. von Renesse and K.-T. Sturm, *Transport inequalities, gradient estimates, entropy, and Ricci curvature*, Comm. Pure Appl. Math. 58 (2005), no. 7, 923–940.